RAPID PROTOTYPING OF DIGITAL SYSTEMS
Second Edition

A Tutorial Approach

RAPID PROTOTYPING OF DIGITAL SYSTEMS
Second Edition

A Tutorial Approach

James O. Hamblen
Georgia Institute of Technology

Michael D. Furman
Georgia Institute of Technology

KLUWER ACADEMIC PUBLISHERS
Boston / Dordrecht / London

Distributors for North, Central and South America:
Kluwer Academic Publishers
101 Philip Drive
Assinippi Park
Norwell, Massachusetts 02061 USA
Telephone (781) 871-6600 Fax (781) 681-9045
E-Mail <kluwer@wkap.com>

Distributors for all other countries:
Kluwer Academic Publishers Group
Distribution Centre
Post Office Box 322
3300 AH Dordrecht, THE NETHERLANDS
Telephone 31 78 6576 000 Fax 31 78 6576 254
E-Mail <services@wkap.nl>

Electronic Services <http://www.wkap.nl>

Library of Congress Cataloging-in-Publication Data

Hamblen, James O., 1954-
Rapid prototyping of digital systems : a tuturial approach / by James O. Hamblen
Michael D. Furman.--2nd ed.
p. cm
Includes index.
ISBN 0-7923-7439-8 (alk. paper)
1. Field programmable gate arrays--Computer-aided design. 2. Programmable array logic. 3. VHDL (Computer hardware description-aided language) I. Furman, Michael D. II. Title

TK7895.G36 H36 2001
621.39'5—dc21

200103837

Printed on acid-free paper. Printed in the United States of America

RAPID PROTOTYPING OF DIGITAL SYSTEMS
SECOND EDITION

Table of Contents

PREFACE

Changes to the Second Edition

The second edition of the text now includes Altera's 10.1 student edition software which adds support for Windows 2000 and designs that are three times larger using the new 70,000 gate UP 1X board. All designs in the book's CDROM have been updated to work with the original UP 1 board or the newer UP 1X board using the new Altera student version software. A coupon is included with the text for purchase of the new UP 1X board. The additional logic and memory in the UP 1X's FLEX 10K70 is useful on larger design projects such as computers and video games.

In addition to the new software, the second edition includes an updated chapter on programmable logic, new robot sensors and projects, optional Verilog examples, and a meta assembler which can be used to develop assembly language programs for the computer designs in Chapters 8 and 13.

Intended Audience

This text is intended to provide an exciting and challenging laboratory component for an undergraduate digital logic design class. The more advanced topics and exercises are also appropriate for consideration at schools that have an upper level course in digital logic or programmable logic. There are a number of excellent texts on digital logic design. For the most part, these texts do not include or fully integrate modern CAD tools, logic simulation, logic synthesis using hardware description languages, design hierarchy, and current generation field programmable logic device (FPLD) technology. The goal of this text is to introduce these topics in the laboratory portion of the course.

Design engineers working in industry may also want to consider this text for a rapid introduction to FPLD technology and logic synthesis using commercial CAD tools, if they have not had previous experience with this new and rapidly evolving technology.

Two tutorials on the Altera CAD tool environment, an overview of programmable logic, and a design library with several easy to use input and output functions were developed for this text to help students get started quickly. Early design examples use schematic capture and library components. VHDL is used for more complex designs after a short introduction to VHDL based synthesis.

The approach used in the text more accurately reflects contemporary practice in industry than the more traditional TTL protoboard based laboratory courses. With modern logic synthesis tools and large FPLDs, more advanced designs are needed to present challenging laboratory projects. Rather than being limited to a few TTL chips that will fit on a small protoboard, designs containing tens of thousands of gates are possible even with student versions of the CAD tools and the UP 1 board. Even student laboratory projects can now implement entire digital systems.

Over the past three years, we have developed a number of interesting and challenging laboratory projects involving serial communications, state machines with video output, video games and graphics, simple computers, keyboard and mouse interfaces, robotics, and pipelined RISC processor cores.

Source files and additional example files are available on the CDROM for all designs presented in the text. The student version of the PC based CAD tool on the CDROM can be freely distributed to students. Students can purchase their own UP 1 board for little more than the price of a contemporary textbook. As an alternative, a few of the low-cost UP 1 or UP 1X boards can be shared among students in a laboratory. Course instructors should contact the Altera University Program for detailed information on obtaining student versions of the CAD tools and UP 1 boards for student laboratories.

Topic Selection and Organization

Chapter 1 is a short CAD tool tutorial that covers design entry, simulation, and hardware implementation using an FPLD. The majority of students can enter the design, simulate, and have the design successfully running on the UP 1 board in less than thirty minutes. After working through the tutorial and becoming familiar with the process, similar designs can be accomplished in less than 10 minutes.

Chapter 2 provides an overview of the UP 1 and UP 1X FPLD development boards. The features of the boards are briefly described. Several tables listing pin connections of various I/O devices serve as an essential reference whenever a hardware design is implemented on the UP 1 board.

Chapter 3 is an introduction to programmable logic technology. The capabilities and internal architectures of the most popular CPLDs and FPGAs are described. These include the MAX 7000 and FLEX 10K family CPLDs used on the UP 1 board, and the Xilinx 4000 family FPGAs.

Chapter 4 is a short CAD tool tutorial that serves as both a hierarchical and sequential design example. A counter is clocked by a pushbutton and the output is displayed in the seven-segment LED's. The design is downloaded to the UP 1 board and some real world timing issues arising with switch contact bounce are resolved. It uses several functions from the UP1core library which greatly simplify use of the UP 1's input and output capabilities.

Chapter 5 describes the available UP1core library I/O functions. The I/O devices include switches, seven-segment LED's, a multiple output clock divider, VGA output, keyboard input, and mouse input.

Chapter 6 is an introduction to the use of VHDL for the synthesis of digital hardware. Rather than a lengthy description of syntax details, models of the commonly used digital hardware devices are developed and presented. Most VHDL textbooks use models developed for simulation only and they frequently use language features not supported in synthesis tools. Our easy to understand synthesis examples were developed and tested using the Altera VHDL CAD tools.

Chapter 7 is a state machine design example. The state machine controls a virtual electric train system simulation with video output generated directly by the CPLD. Using track sensor input, students must control two trains and three track switches to avoid collisions.

Chapter 8 develops a model of a simple computer. The fetch, decode, and execute cycle is introduced and a brief model of the computer is developed using VHDL. A short assembly language program can be entered in the FPLD's internal memory and executed in the simulator.

Chapter 9 describes how to design an FPLD-based digital system to output VGA video. Numerous design examples are presented containing video with both text and graphics. Fundamental design issues in writing simple video games and graphics using the UP 1 board are examined.

Chapter 10 describes the PS/2 keyboard operation and presents interface examples for integration in designs on the UP 1 board. Keyboard scan code tables, commands, status codes, and the serial communications protocol are included. VHDL code for a keyboard interface is also presented.

Chapter 11 describes the PS/2 mouse operation and presents interface examples for integration in designs on the UP 1 board. Commands, data packet formats, and PS/2 serial communications protocols are included.

Chapter 12 develops a design for an adaptable mobile robot using the UP 1 board. Servo motors and several sensor technologies for a low cost mobile robot are described. A sample servo driver design is presented. Commercially available parts to construct the robot described in Chapter 12 can be obtained for around $50. Several robots can be built for use in the laboratory. Students with their own UP 1 board may choose to build their own robot following the detailed instructions found in section 12.6.

Chapter 13 describes a single clock cycle model of the MIPS RISC processor based on the hardware implementation presented in the widely used Patterson and Hennessy textbook, *Computer Organization and Design The Hardware/Software Interface*. Laboratory exercises that add new instructions, features, and pipelining are included at the end of the chapter.

We anticipate that many schools will still choose to begin with TTL designs on a small protoboard for the first few labs. The first chapter can also be started at this time since only OR and a NOT logic functions are used to introduce the CAD tool environment. The CAD tool can also be used for simulation of TTL labs, since an extensive TTL parts library is included with the student version.

Even though VHDL is a complex language, we have found after several years of experimentation that students can write VHDL to synthesize hardware designs after a short overview with a few basic hardware design examples. The use of VHDL templates and online help files in the CAD tool makes this process easier. After the initial experience with VHDL synthesis, students dislike the use of schematic capture on larger designs since it can be very time consuming. Experience in industry has been much the same since huge productivity gains have been achieved using HDL based synthesis tools for application specific integrated circuits (ASICs) and FPLDs.

Most digital logic classes include a simple computer design such as the one presented in Chapter 8 or a RISC processor such as the one presented in Chapter 13. If this is not covered in the first digital logic course, it could be used as a lab component for a subsequent computer architecture class.

A typical quarter or semester length course could not cover all of the topics presented. The material presented in Chapters 7 through 13 can be used on a selective basis. The keyboard and mouse are supported by UP1core library functions, and the material presented in Chapters 10 and 11 is not required to use these library functions for keyboard or mouse input. Information on video generation, PS/2 keyboard, and mouse interfacing is not readily available elsewhere.

A video game based on the material in Chapter 9 can serve as the basis for a team design project. For a final team design project, we use robots with sensors from chapter 12 that are controlled by the simple computer in chapter 8. The meta assembler in Appendix D is used to assemble the programs. Our students really enjoyed working with the robot described in Chapter 12, and it presents almost infinite possibilities for an exciting design competition.

Software and Hardware Packages

The new student version of Altera's MAX+PLUS II CAD tool is included with this book. UP 1 and UP 1X boards are available from Altera at special student pricing. A board can be shared among several students in a lab, or some students may wish to purchase their own board. Details and suggestions for additional cables and power supplies that may be required for a laboratory setup can be found in Section 2.5. Source files for all designs presented in the text are available on the CDROM.

Additional Web Material and Resources

There is a web site for the text with additional course materials, slides, text errata, and software updates at:

http://www.ece.gatech.edu/users/hamblen/book/bookse.htm

Acknowledgments

Over a thousand students and several hundred teaching assistants have contributed to this work during the past two years. In particular, we would like to acknowledge Doug McAlister, Michael Sugg, Jurgen Vogel, Tyson Hall, Greg Ruhl, Eric Van Heest, and Mitch Kispet for their help in testing and developing several of the laboratory assignments and tools. Joe Hanson, Tawfiq Mossadak, and Eric Shiflet at Altera provided software, hardware, helpful advice, and encouragement.

RAPID PROTOTYPING OF DIGITAL SYSTEMS

SECOND EDITION

A Tutorial Approach:

- Applications
- Design
- Synthesis
- Simulation
- Implementation

JAMES O. HAMBLEN AND MICHAEL D. FURMAN
SCHOOL OF ELECTRICAL AND COMPUTER ENGINEERING
GEORGIA INSTITUTE OF TECHNOLOGY
ATLANTA, GEORGIA

CHAPTER 1

Tutorial I: The 15-Minute Design

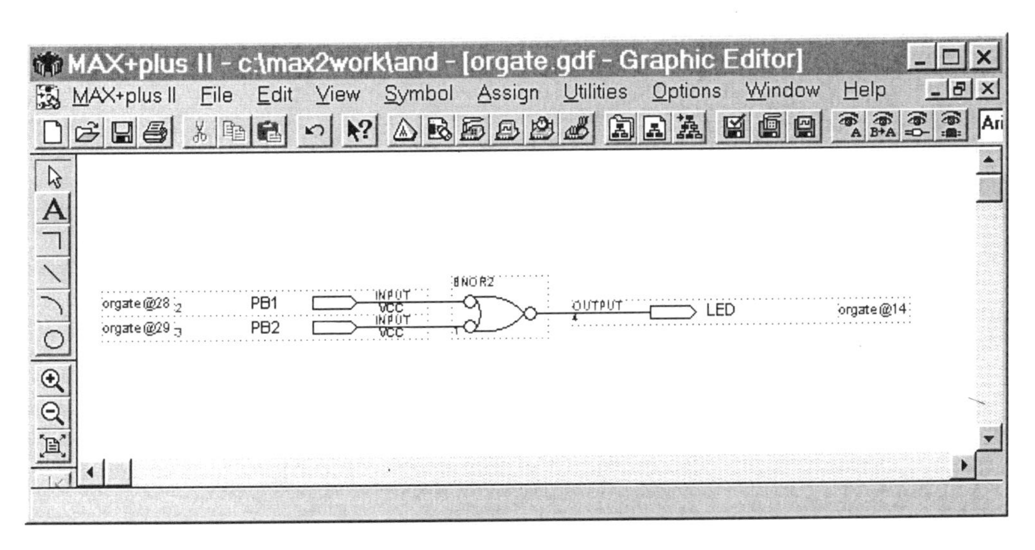

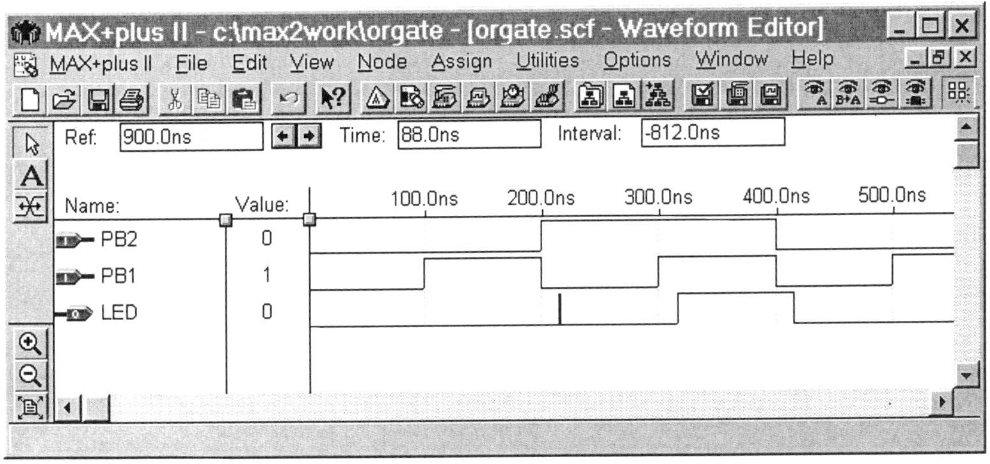

1 Tutorial I: The 15 Minute Design

The purpose of this tutorial is to introduce the user to the Altera CAD tools and the University Program (UP 1, UP 1X, or UP 2) Development Board in the shortest possible time. The format is an aggressive introduction to schematic, VHDL, and Verilog entry for those who want to get started quickly. The approach is tutorial and utilizes a path that is similar to most digital design processes.

Once completed, you will understand and be able to:

- Navigate the Altera schematic entry environment,
- Simulate, debug, and test your designs,
- Generate and verify timing characteristics, and
- Download and run your design on a UP 1, UP 1X, or UP 2 board.

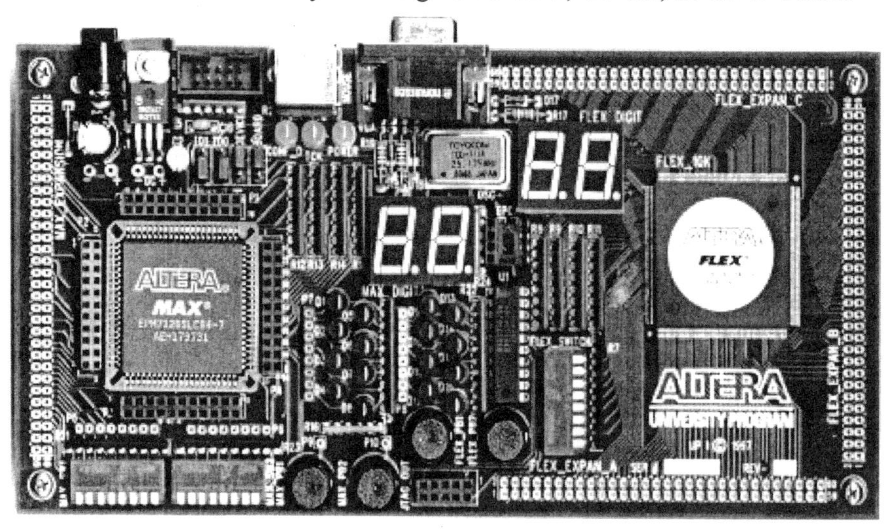

Figure 1.1 The Altera UP 1 CPLD development board.

In this tutorial, an OR function will be demonstrated to provide an introduction to the ALTERA MAX+PLUS II CAD tools. After simulation, the design will then be used to program a complex programmable logic device (CPLD) on a UP 1, UP 1X, or UP 2 development board. Altera's newest board, the UP 2, is virtually identical to the UP 1X. If you have a UP 2 board, you should always follow the instructions and procedures outlined in the CDROM and text for a UP 1X board. The inputs to the OR logic will be two pushbuttons and the output will be displayed using a light emitting diode (LED). Both the pushbuttons and the LED are part of the development board, and no external wiring is required. Of course, any actual design will be more complex, but the objective here is to quickly understand the capabilities and flow of the design tools with minimal effort and time.

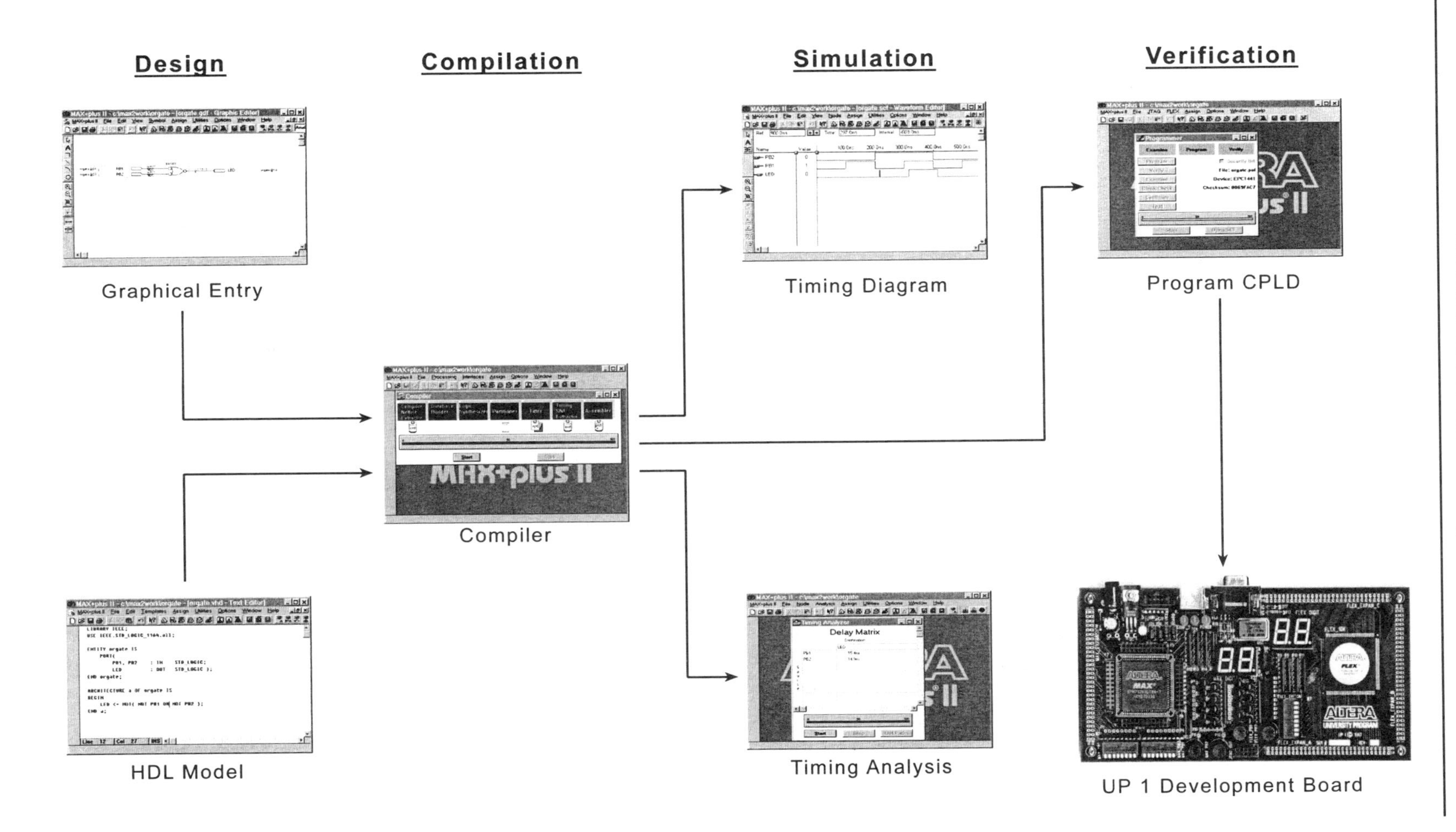

Figure 1.2 Design process for schematic or HDL entry.

Granted, all this may not be accomplished in just 15 minutes; however, the skills acquired from this demonstration tutorial will enable the first-time user to duplicate similar designs in less time than that!

INSTALL THE ALTERA MAX+PLUS II STUDENT VERSION USING THE CDROM AND OBTAIN A STUDENT AUTHORIZATION CODE VIA EMAIL FROM ALTERA. CHECK FOR ALTERA STUDENT VERSION SOFTWARE UPDATES AT WWW.ALTERA.COM

Designs can be entered via schematic capture or by using a hardware description language (HDL) such as VHDL, AHDL, or Verilog. It is also possible to combine blocks with different entry methods into a single design. As seen in Figure 1.2, tools can then be used to simulate, calculate timing delays, synthesize logic, and program a hardware implementation of the design into a CPLD.

The Board

The board that will be used, the Altera UP 1, is actually two CPLD development boards in one. The Altera MAX 7128S series chip on the left side of the board has a gate count of over 2,500, while the FLEX 10K20 series chip on the right of the board has over 20,000 gates. Although the following tutorial can be done with either chip, the higher density FLEX EPF10K20 will be used, since the pushbuttons and LEDs can be utilized without having to add additional jumper wires. An exercise at the end of the tutorial contains suggestions for performing the tutorial on the smaller MAX chip.

The Pushbuttons

The FLEX input pushbuttons PB1 and PB2 are connected directly to the FLEX chip at pins 28 and 29 respectively. Each input is tied High with a pull-up resistor and pulled Low when the respective pushbutton is pressed. One needs to remember that when using the on-board pushbuttons, this "active low" condition ties zero volts to the input when the button is pressed and five volts to the input when not pressed. See Figure 1.3.

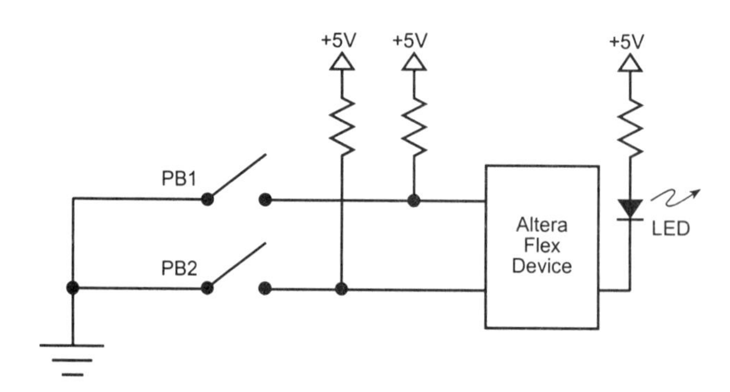

Figure 1.3 Connections between the pushbuttons, the LEDs, and the Altera FLEX device.

The LED Outputs

The seven-segment LEDs can be used for monitoring outputs. These LEDs are also pre-wired to the FLEX chip with a pull-up resistor as illustrated in Figure 1.3. This configuration allows the external resistor to control the amount of current through the LED; however, it also requires the FLEX chip to output a Low signal to turn on the LED. (Students regularly forget this point and have a fully working project with an inverted pattern on the LEDs.)

Figure 1.4 shows FLEX pin number 14 hard wired to the seven-segment LED decimal point. (A full listing of the other segments and the pinouts can be found in Appendix C.2. For the purposes of this tutorial, only the decimal point will be used for output.)

Figure 1.4 FLEX chip pin connection to seven-segment display decimal point.

The Problem Definition

To illustrate the capabilities of the software in the simplest terms, we will be building a circuit that turns on the decimal-point LED when one OR the other pushbutton is pushed. In simple logic:

$$LED = PB1 + PB2$$

At first, this may seem too simple; however, the active low inputs and outputs add just enough complication to illustrate some of the more common errors, and it provides an opportunity to compare some of the different syntax features of VHDL and Verilog. (Students needing an exercise in DeMorgan's Law will also find these exercises particularly enlightening.)

We will first build this circuit with the graphical editor and then implement it in VHDL and Verilog. As you work through the tutorial, note how the design entry method is relatively independent of the compile, timing, and simulation steps.

Resolving the Active Low Signals

Since the pushbuttons generate inverted signals and the LEDs require inverted signals to turn on, we could build an OR logic circuit using the layout in Figure 1.5a. Recalling that a bubble on a gate input or output indicates inversion, careful examination shows that the two circuits in Figure 1.5 are functionally equivalent; however, the circuit in Figure 1.5a uses more gates. We will therefore use the single circuit illustrated in Figure 1.5b.

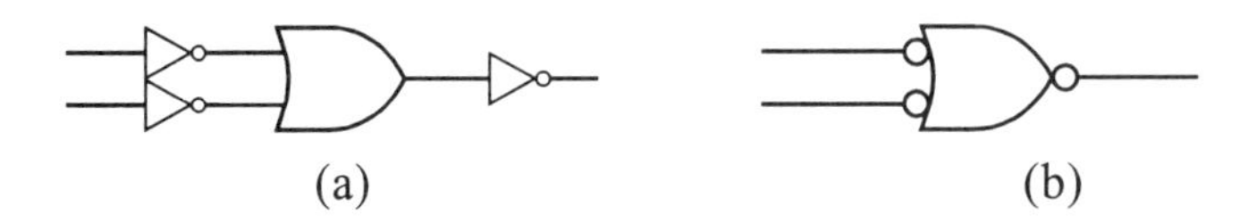

Figure 1.5a and 1.5b. Equivalent circuits for ORing active low inputs and outputs.

This form of the OR function is known as a "negative-logic OR." (In Exercise 1 at the end of the chapter, this circuit will be compared with its DeMorgan's equivalent, the "positive-logic AND.")

1.1 Design Entry using the Graphic Editor

Examine the CAD tool overview diagram in Figure 1.2. The initial path in this section will be from schematic capture (Graphical Entry) to downloading the design to the UP 1 board. On the way, we will pass through some of the nuances of the Compiler along with setting up and controlling a simulation. Later, after having actually tested the design, a Timing Analysis will be run on the design. Although relatively short, each step is carefully illustrated and explained.

Establishing Graphics (Schematic) as the Input Format

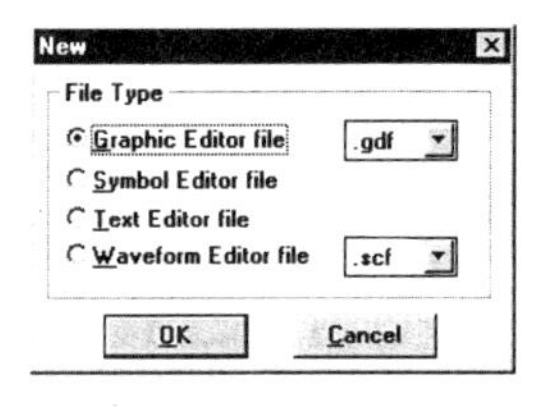

Start the MAXPLUS program. Choose **File ⇨ New,** and a popup menu will appear. Select **Graphic Editor,** then click **OK**. This will create a blank schematic worksheet – a graphics display file (*.gdf file). Note that the top line menu options are context sensitive and change as different tools are selected.

Select the Device to be Used

Next, select **Assign ⇨ Device**. Under **Device Family** select the **FLEX 10K** device family. For UP 1 boards, under **Devices** select **EPF10K20RC240-4** as the device (-*X* is the speed grade of the chip. Most UP 1 boards contain a –4 chip.) Check the device number and speed grade printed on the FLEX chip on the right side of the board just above the large Altera University Program Logo. If you don't see the –4 part listed in the menu, uncheck the show fastest devices only box. Click **OK**.

If you are using the new larger FLEX 10K70 70,000 gate UP 1X or UP 2 board, you will need to select a **EPF10K70RC240-*X*** device for all of your designs. This will need to changed from the default value with **Assign ⇨ Device** in each of the book's CDROM designs before they are compiled. The design cannot be downloaded to the board if the device number was incorrect when the design was compiled.

Enter and Place the OR Symbol in Your Schematic

Choose **Symbol ⇨ Enter Symbol** then on the popup menu select **OK.** This will tell the software to automatically place the selected device in the center. In the symbol library window, double click the library path that ends with ...**\prim**. Scroll down the list of symbols and double click **BNOR2**. An OR gate with inverted inputs and outputs should appear on the schematic. (The naming convention is **B**-bubbled **NOR** with **2** inputs. Although considered to be a NOR with active low inputs, it is fundamentally an OR gate with active low inputs and output.)

> TO GET CONTEXT SENSITIVE HELP, CLICK ON THE ICON, AND THEN CLICK ON THE **BNOR2** SYMBOL. AT ANY POINT IN THE TUTORIAL, EXTENSIVE ONLINE HELP IS ALWAYS AVAILABLE. TO SEARCH BY TOPIC OR KEYWORD SELECT THE **HELP** MENU AND FOLLOW THE INSTRUCTIONS THERE.

Assigning the Output Pin for Your Schematic

Select **Symbol ⇨ Enter Symbol** and click **OK.** Again select the library path that ends with ...**\prim**. Under **Symbol Files** double click on **output**. Using the mouse and the left mouse button, drag the output symbol to the right of the **BNOR2** symbol leaving space between them – they will be connected later.

Assigning the Input Pins for Your Schematic

Find and place two **input** symbols to the left of the **BNOR2** symbol in the same way that you just selected the **output** symbol. (Try using the right mouse button and **Enter Symbol** instead of the **Symbol ⇨ Enter Symbol** pull-down. Another hint: Once selected, a symbol can be copied and pasted using the right mouse button.)

Connecting the Signal Lines

Using the mouse, move to the end of one of the wires. A cross-symbol mouse cursor should appear when the mouse is near a wire. Drag the end of the wire to the point that you want to connect. If you need to delete a wire, click on it – the wire should turn red when selected. Hit the delete key to remove it. You can also use the right mouse key and select delete. Connect the other wires using the same process so that the diagram looks something like Figure 1.6.

Enter the PIN Names

Right click on the first **INPUT** symbol. It will be outlined in red and a menu will appear. Select **Edit Pin Name**. Type **PB1** for the pin name. Name the other input pin **PB2** and the output pin for the **LED** in a similar fashion.

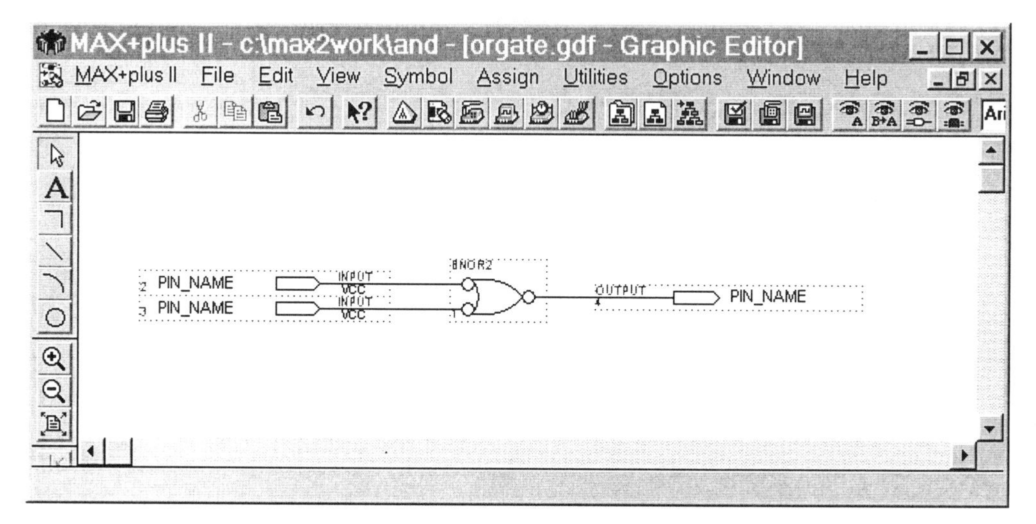

Figure 1.6 Active low OR-gate schematic example with pinouts.

Assign the PIN Numbers to Connect the Pushbuttons and the LED

Since the FLEX chip on the UP 1 board is prewired to the pushbuttons and the LEDs, we need to look up the pin numbers and designate them in our design. The relevant information in Table 1.1 is from the documentation on the pinouts for the UP 1 board. (See Table 2.4.)

Table 1.1 Hardwired connections on the FLEX chip for the design.

Device	Pin Number Connections
PB1	28
PB2	29
LED Decimal Point On the Left Seven-Segment Display	14

On your diagram select the PB1 input symbol. Right click and select **Assign ⇨ Pin/Location/Chip**. Under **Chip Resource,** select **Pin**. (If the option to select the pin is unavailable, you need to **Cancel** this menu, go back and select **Assign ⇨ Device,** and make sure that your device is selected correctly.) Next to **Pin**, select the down arrow and scroll down to select pin 28 or just enter 28 in the space provided. Repeat this process assigning pin 29 to PB2 and pin 14 to LED. Device and pin information is stored in the project's *.acf file. Caution: Be sure to use unique names for different pins. Pins and wires with the same name are automatically connected without a visible wire showing up on the schematic diagram.

Saving Your Schematic

Select **File ⇨ Save As** and use the max2work directory or a new sub-directory for your project directory. Name the file **ORGATE**. (The software requires the working directory to not be the root directory on some network drives.) Note that after saving your schematic, the pin numbers change to reflect the name of the file. Throughout the remainder of this tutorial you should always refer to the same project directory path when saving or opening files. The schematic should now look like Figure 1.7. A number of other files are automatically created by the MAX+PLUS II tools and maintained in your project directory.

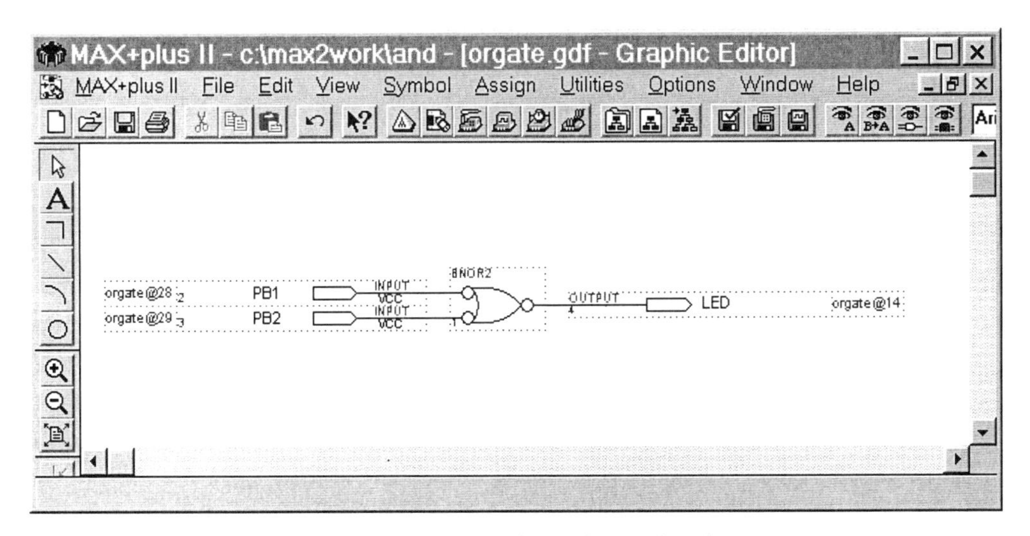

Figure 1.7 Active low OR-gate schematic with pin numbers assigned.

1.2 Compiling the Design

Compiling your design checks for syntax errors, synthesizes the logic design, produces timing information for simulation, fits the design on the selected CPLD, and generates the file required to download the program. After any changes are made to the design files or pin assignments, the project should always be re-compiled prior to simulation or downloading.

Set Project to Current File and Compile

Select **File ⇨ Project ⇨ Set Project to Current File.**

Compile by selecting **File ⇨ Project ⇨ Save and Compile**. A window with the modular compiler, as illustrated in Figure 1.8, will monitor compiling, warnings, and errors.

Checking for Compile Warnings and Errors

The project should compile with **0 Errors**. If the message states, "Project Compilation was Successful," then you have not made an error. Warnings about forcing pin assignments can be ignored. (The compiler wants to select pins based on the best performance for internal timing and routing. Since the pins for the pushbuttons and the LED are pre-wired on the UP 1 board, their assignment cannot be left up to the compiler.)

Examining the Report File

View the orgate.rpt file by double clicking on the rpt icon below the fitter block in the compile window. After compilation, a *.rpt file is created that contains information on device utilization, pin assignments, and a copy of the netlist. After reviewing this file, close it and return to the circuit with **File ⇨ Close**.

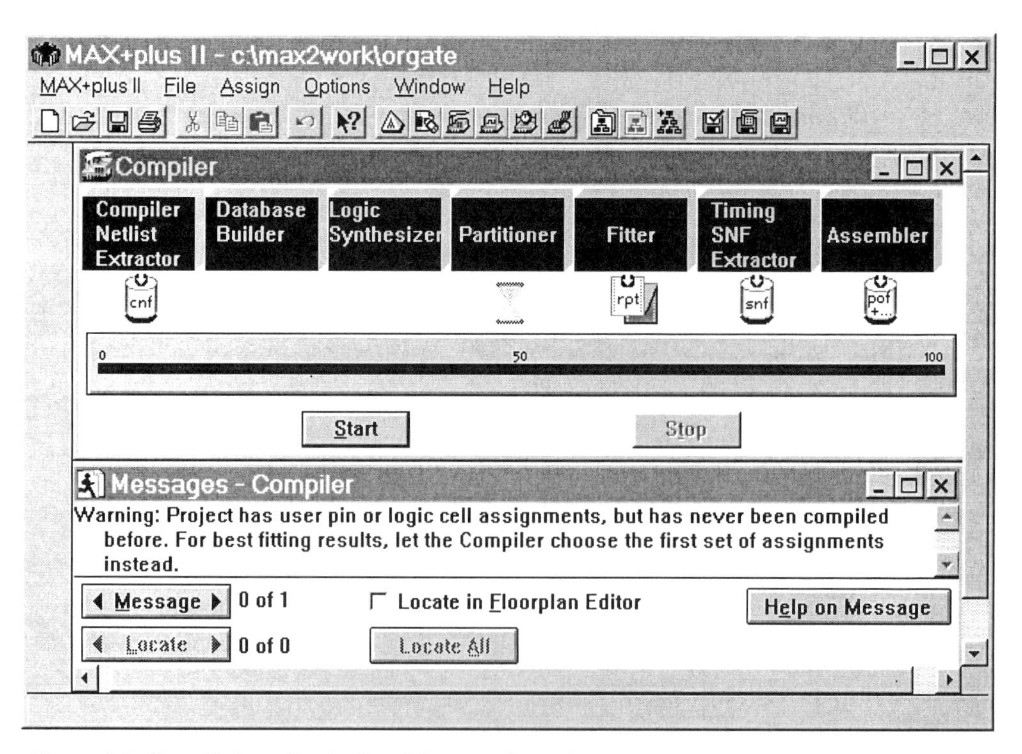

Figure 1.8 Compilation of active low OR-gate schematic.

1.3 Simulation of the Design

For complex designs, the design would normally be simulated prior to downloading to a CPLD. Although the OR example is straightforward, we will take you through the steps to illustrate the simulation of the circuit.

Set Up the Simulation Traces

Choose **File ⇨ New**, select **Waveform Editor File**, and then from the popup window click **OK**. The waveform window should be displayed. Select **Node ⇨ Enter Nodes from SNF**. (A SNF is a Simulator Netlist File. It contains timing information about your circuit that allows the software to produce a timing diagram output display.) Click on the **LIST** button. PB1, PB2 and LED should appear as trace values in the window. Then click on the center => button and click **OK**.

Generate Test Vectors for Simulation

A simulation requires external input data or "stimulus" data to test the circuit. Since the PB1 and PB2 signals have not been set, the simulator sets them to a default of zero. The 'X' on the LED trace indicates that the simulator has not yet been run. (If the simulator *has* been run and you still get an 'X,' then the simulator was unable to determine the output condition.)

Right click on **PB1**. The PB1 trace will be highlighted. Select **Overwrite ⇨ Count Value...** and click **OK**. An alternating pattern of Highs and Lows should appear in the PB1 trace. Next, select **View ⇨ Time Range...** and set the **From** and **To** range to 0.0ns and 500.0ns respectively.

Right click on **PB2**. Select **Overwrite ⇨ Count Value...**, and change the entry for **Multiplied By** from **1** to **2,** and click **OK**. **PB2** should now be an alternating pattern of ones and zeros but at half the frequency of PB1. (Other useful options in the **Overwrite** menu will generate a clock and set a signal High or Low. It is also possible to highlight a portion of a signal trace with the mouse and set it High or Low manually.)

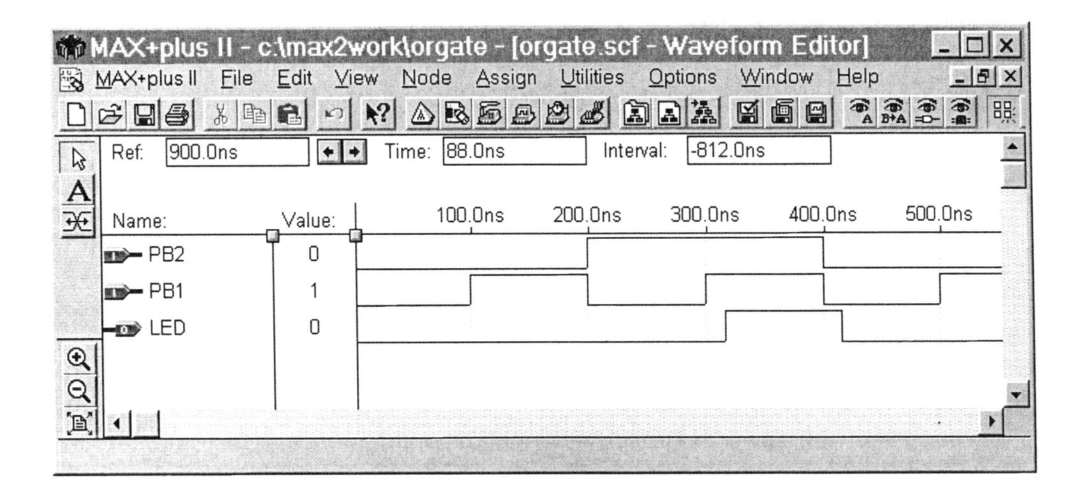

Figure 1.9 Active low OR-gate timing simulation.

Performing the Simulation with Your Timing Diagram

Select **File ⇨ Project ⇨ Save, Compile, and Simulate**. Click **OK**. The simulation should run and the waveform for LED should now appear as seen in Figure 1.9. Note that the simulation includes the actual device timing delays through the device and that it takes around 20ns (ns = 10^{-9} sec.) for the output to reflect the new inputs. Verify that the LED output is Low only when either PB1 OR PB2 inputs are Low.

1.4 Downloading Your Design to the UP 1 or UP 1X Board

Hooking Up the UP 1 Board to the Computer

Plug the Byteblaster* cable into the UP 1 or UP 1X board and attach the other end to the parallel port on a PC. If you have not done so already, make sure that the PC's BIOS settting for the printer port is ECP or EPP mode. Using a 9V AC to DC wall transformer or another 7 to 9V DC power source, attach power to the DC power connector (DC_IN) located on the upper left-hand corner of the UP 1 board. When properly powered, one of the green LEDs on the board should light up. If a DC wall transformer is not available, Section 2.5 lists suggested sources for a low cost AC to DC wall transformer with the proper connector, voltage, amperage, and polarity.

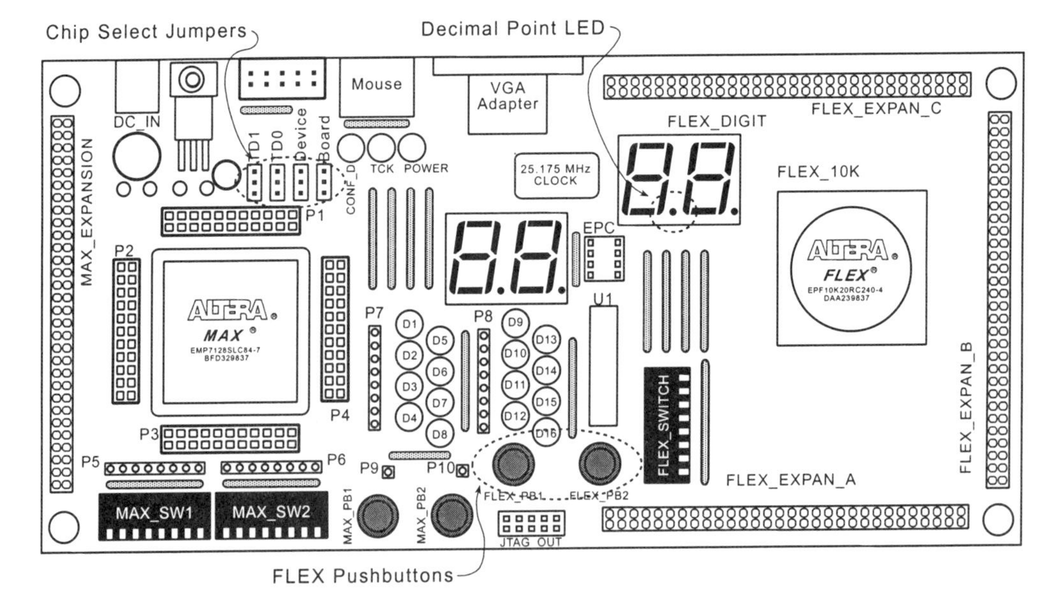

Figure 1.10 ALTERA UP 1/UP 1X board with jumper settings and PB1, PB2, and LED locations.

Verify that the device jumpers are set for the FLEX chip as shown in Table 1.2. The locations of the pushbuttons, PB1 and PB2, and the LED decimal point are also highlighted in Figure 1.10. (Note that for the MAX EPM7128 chip, the jumper pins are all set to the top position as indicated in Table 1.2.)

Table 1.2 Jumper settings for downloading to the MAX and FLEX devices.

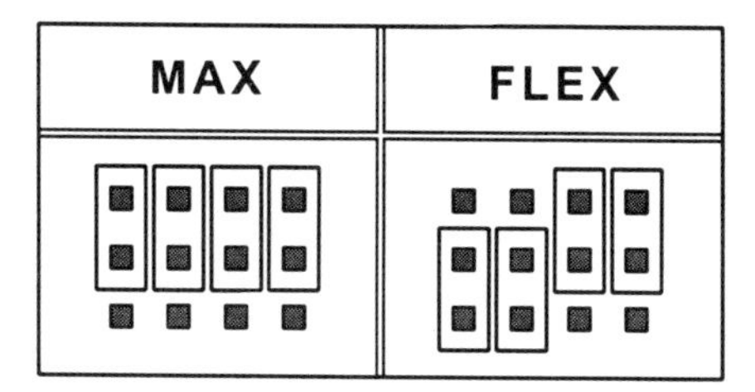

Preparing for Downloading

After checking to make sure that the cables and jumpers are hooked up properly, you are ready to download the compiled circuit to the UP 1 board. Select **MAX+PLUS II ⇨ Programmer**. From **Options ⇨ Hardware ⇨ Setup**, select the proper output port: normally **LPT1**. (If a window comes up that displays, "**No Hardware**", use the pull-down to change "**No Hardware**" to "**Byteblaster**". If this still doesn't correct the problem, then there is something else wrong with the setup. Go back to the beginning of this section and check each connection carefully.)

JTAG Setup

THIS NEEDS TO BE DONE EACH TIME THAT THE DESIGN FILENAME IS CHANGED.

Select **JTAG ⇨ Multi-Device JTAG chain** so that a checkmark appears to the left of the option. Next select **JTAG ⇨ Multi-Device JTAG Setup**. Hit the **Select Programming File** button and select **orgate.sof**. Click **OK**, then click the **ADD** button and the new filename should move into the list in the inner window. Delete any other entries that may be in this window.

Select the **Detect JTAG Chain Info** button. The system should respond with "JTAG chain information confirmed by hardware check." If not, double check cables, power, jumpers, and make sure you have the correct file name and chip listed in the inner window. Click **OK** to exit the **JTAG Multi-Device Setup** window.

Final Steps to Download

Make sure that the **Device Name** has changed to **EP10K20** or **EPF10K20RC240** or **EPF10K70RC240** for the UP 1X (depending on the UP board and MAX+PLUS II version that you are running). If it does not display the correct device, then return to your schematic, assign the correct device (See section 1.1.), recompile, and try again.

The **Configure** button in the programming window should now be highlighted. Click on the **Configure** button to download to the UP 1 board. Just a few seconds are required to download. If download is successful, a window with

Configuration Complete is displayed – click **OK**. (If the **Configure** button is not highlighted, try **Options ⇨Hardware Setup** from the pull-down window. Confirm the port settings and click **OK**. Also, confirm that the JTAG setup dialog information is correct. If you still have problems confirm that the printer port BIOS settings use ECP or EPP mode.)

Testing Your Design

The locations of PB1, PB2, and the decimal LED are indicated in Figure 1.10. After downloading your program to the UP 1 board, the two rightmost seven-segment displays should light up. Since the output of the OR gate is driving the decimal LED signal on the left digit of the two seven-segment displays, it should be off. Since the buttons are active low, and the **BNOR2** gate also has active low inputs and output, hitting either button should turn on the LED.

Congratulations! You have just entered, compiled, simulated, and downloaded a design to a CPLD device, and verified its operation.

1.5 The 10 Minute VHDL Entry Tutorial

As an alternative to schematic capture, a hardware description language such as VHDL, Verilog, or AHDL can be used. In large designs, these languages greatly increase productivity and reduce design cycle time. Logic minimization and synthesis to a netlist are automatically performed by the compiler and synthesis tools. (A netlist is a textual representation of a schematic.) As an example, to perform addition, the VHDL statement:

A <= B + C;

will automatically generate an addition logic circuit with the correct number of bits to generate the new value of A.

Using the OR-gate design from the Schematic Entry Tutorial, we will now create the same circuit using VHDL.

PRIOR KNOWLEDGE OF VHDL IS NOT NEEDED TO COMPLETE THIS TUTORIAL.

Using a Template to Begin the Entry Process

Choose **File ⇨ New**, select **Text Editor File** and **OK**. Place the cursor within the text area, right click the mouse, and select **VHDL Template**. (Note the different prewritten templates. These are provided to expedite the entry of VHDL.) Select **ENTITY** – this declaration is the one you will generally start with since it also sets up the input and output declarations. The template for the ENTITY declaration appears highlighted in the text-editor window. Click the mouse to unhighlight the text.

Saving the File to Set the Editor Convention

Select **File ⇨ Save As**. Change the **Automatic Extension** to **.vhd** (VHDL) and save the file as **orgate.vhd** – click **OK**. Now that the editor knows that it is a VHDL source file, the text will appear in different context-sensitive colors. VHDL keywords appear in blue and strings in green. The coloring information should be used to detect syntax errors while still in the text editor.

Replacing Comments in the VHDL Code

The entire string indicating the position of the entity name, **__entity_name**, should be set to the name used for the filename – in this case, **orgate**. There are two occurrences of **__entity_name** in the text. Find and change both accordingly.

Declaring the I/O Pins

The input and output pins, PB1, PB2 and LED need to be specified in the PORT declaration. Since there are no input vectors, bi-directional I/O pins, or GENERIC declarations in this design, remove all of these lines. The source file should look like Figure 1.11.

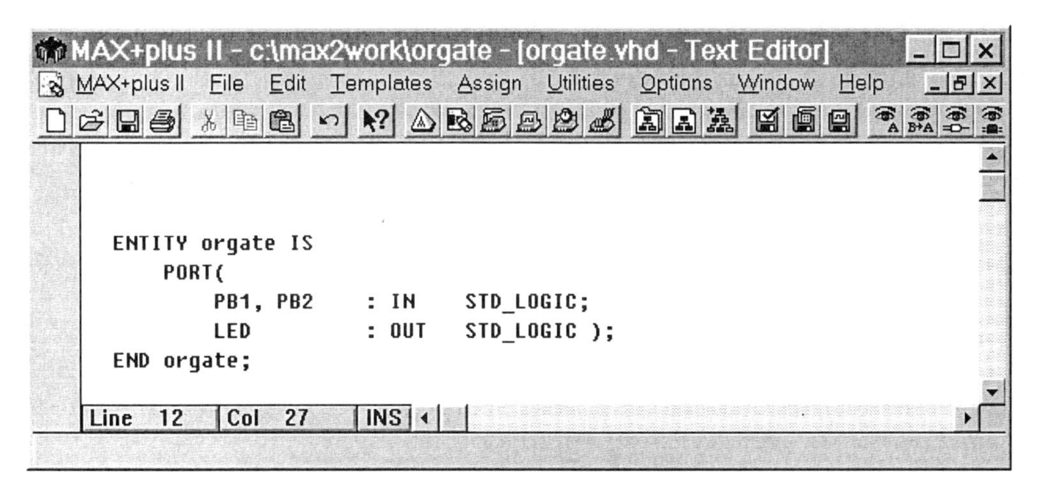

Figure 1.11 VHDL Entity declaration text.

Setting up the Architecture Body

Click the mouse at the bottom of the text field. (We will be inserting another template here.) Following the earlier procedure for selecting a VHDL template (start with a right click), select an **Architecture Body**. (The **Architecture Body** specifies the internal logic of the design.) The syntax for the **Architecture Body** appears in the text window after the other text. (You can now see why the template is left highlighted – had you not placed your cursor first, text would have appeared at your last cursor position. If you do misplace the template, hitting the delete key removes the highlighted text.)

Editing the Architecture Body

Change the entity name in the **ARCHITECTURE** statement to **orgate**. Template lines with a "--" preceding a comment, need to be edited for each particular design. Delete the two signal declaration lines since this simple design does not require internal signals. Delete the remaining comment lines that start with "--", and insert **LED <= NOT (NOT PB1 OR NOT PB2)** as a single line. (This line contains a deliberate syntax error that will be detected and fixed later.)

Insert the following two lines at the beginning of the text file to define the libraries for the STD_LOGIC data type.

```
LIBRARY IEEE;
USE IEEE.STD_LOGIC_1164.all;
```

This is the preferred data type for bits in VHDL. The file should now appear as in Figure 1.12.

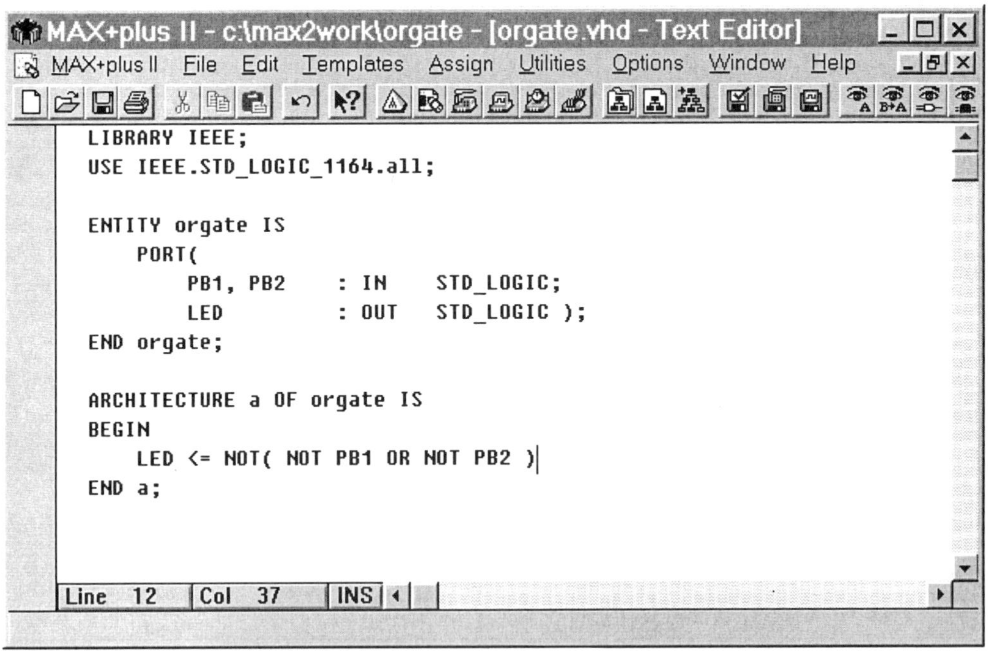

MAX+plus II - c:\max2work\orgate - [orgate.vhd - Text Editor]

MAX+plus II File Edit Templates Assign Utilities Options Window Help

```
LIBRARY IEEE;
USE IEEE.STD_LOGIC_1164.all;

ENTITY orgate IS
    PORT(
        PB1, PB2    : IN    STD_LOGIC;
        LED         : OUT   STD_LOGIC );
END orgate;

ARCHITECTURE a OF orgate IS
BEGIN
    LED <= NOT( NOT PB1 OR NOT PB2 )
END a;
```

Line 12 Col 37 INS

Figure 1.12 VHDL OR-gate model (with syntax error).

Before You Compile

Before you compile the VHDL code, the device type and pin numbers need to be assigned. Select **Assign ⇨ Device**. Under **Device Family** select the **FLEX 10K** family. Under **Devices** select the **EPF10K20RC240-*x*** chip. If you are using the new larger FLEX 10K70 70,000 gate UP 1X board you will need to select a **EPF10K70RC240-*X*** device for all of your designs. (-*x* is the chips speed grade that can vary with the different versions of UP 1 board. –*x* should be the same as the markings on the FLEX chip on the UP 1 board for accurate simulation times.) Many UP 1 boards have the –4 part. To find the correct speed grade, uncheck the show only fastest speed grade box in the **Assign Device** window).

Next, select **Assign ⇨Pin/Location/Chip**. Select **Pin** and scroll down to assign pin **28** to PB1. If your pins are already defined from the earlier Schematic Entry Tutorial, just confirm the pin assignments. Repeat this process and assign or confirm that pin 29 is connected to PB2 and that pin 14 is connected to the decimal LED. At this point, VHDL code is generally ready to be compiled, simulated, and downloaded to the board using steps identical to those used earlier in the schematic entry method. Once pin assignments are made, they are stored in the project's *.acf file.

1.6 Compiling the VHDL Design

The Compile process checks for syntax errors, synthesizes the logic design, produces timing information for simulation, fits the design on the CPLD, and generates the file required to program the CPLD. After any changes are made to the design files or pin assignments, the project should always be re-compiled prior to simulation or programming. The compile is a seven-step process.

Select **File ⇨ Project ⇨ Set Project to Current File**. Next, compile by selecting **File ⇨ Project ⇨ Save and Compile**.

Checking for Compile Warnings and Errors

The project should compile with **1 Error**. After compiling the VHDL code, a window indicating an error should appear. The result should look something like Figure 1.13.

Double click on the red error line and note that the cursor is placed in the editor either on or after the line missing the semicolon(;). VHDL statements should end with a semicolon. Add the semicolon to the end of the line so that it is now reads:

LED <= NOT (NOT PB1 OR NOT PB2);

Now, recompile, and you should have no errors.

1.7 The 10 Minute Verilog Entry Tutorial

Verilog is another widely used hardware description language (HDL). Verilog and VHDL have roughly the same capabilites. VHDL is based on a PASCAL style syntax and Verilog is based on the C language. In large designs, HDLs greatly increase productivity and reduce design cycle time. Logic minimization and synthesis are automatically performed by the compiler and synthesis tools.

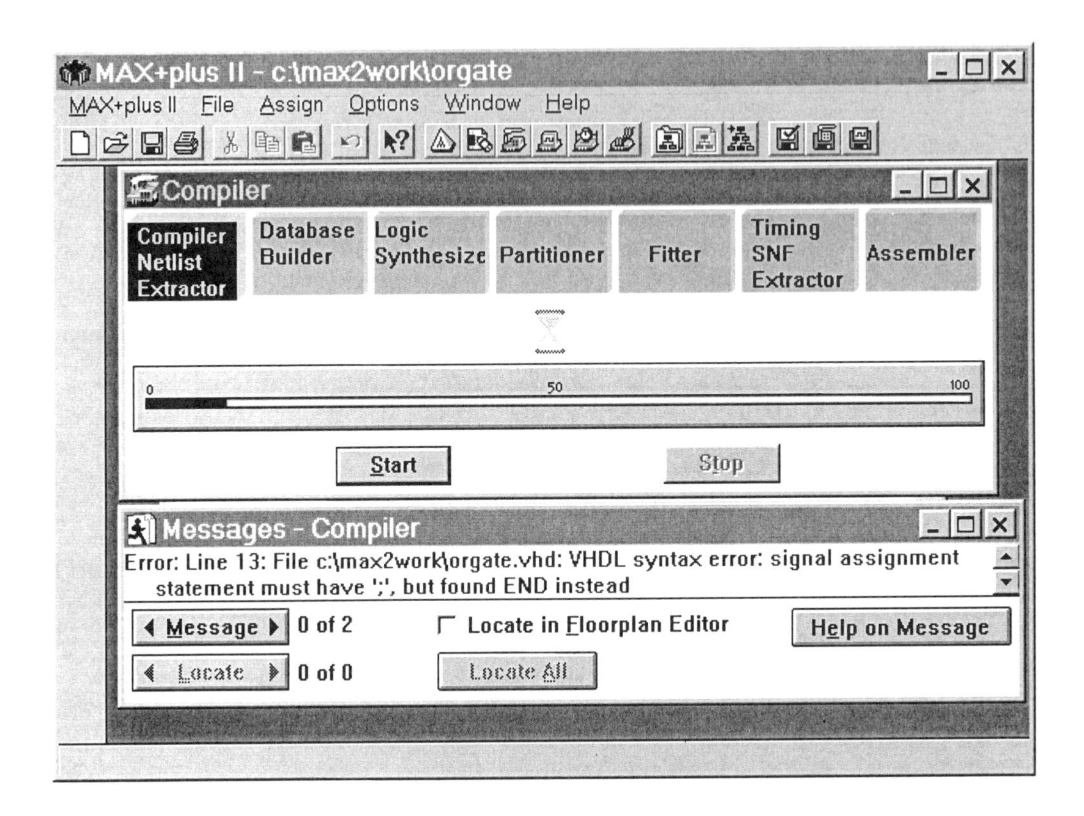

Figure 1.13 VHDL compilation with a syntax error.

Just like the previous VHDL example, to perform addition, the Verilog statement:

A = B + C;

will automatically generate an addition logic circuit with the correct number of bits to generate the new value of A.

Using the OR-gate design from the Schematic Entry Tutorial and the VHDL Tutorial, we will now create the same circuit in Verilog.

PRIOR KNOWLEDGE OF VERILOG IS NOT NEEDED TO COMPLETE THIS TUTORIAL.

Using a Template to Begin the Entry Process

Choose **File ⇨ New**, select **Text Editor File** and **OK**. Place the cursor within the text area, right click the mouse, and select **Verilog Template**. (Note the different prewritten templates. These are built to expedite the entry of Verilog.) Select **Module Declaration** – this declaration is the one you will generally start with since it also sets up the input and output declarations. The template for the module declaration appears in the Text editor window highlighted. Click the mouse to unhighlight the text.

Saving the File to Set the Editor Convention

Select **File ⇨ Save As**. Change the **Automatic Extension** to **.v** (Verilog) and save the file as **orgate.v** – click **OK**. Now that the editor knows that it is a Verilog source file, the text will appear in different context-sensitive colors. Verilog keywords appear in blue and strings in green. The coloring information should be used to detect syntax errors while still in the text editor.

Replacing Comments in the Verilog Code

The entire string indicating the position of the entity name, **__module_name**, should be set to the name used for the filename – in this case, **orgate**. There is one occurrence of **__module_name** in the text. Find and change it accordingly. Lines starting with // are comments and these will need to be replaced with the appropriate Verilog code.

Declaring the I/O Pins

Move the cursor just after the line starting with "// PORT ..." and insert the Port Declaration template. The input pins, PB1, PB2 and the output pin LED need to be specified in the arguments of the Module statement and Port declaration. Since there are no inout pins, wire or integer declarations, or Always statements in this design, remove all of these lines. The source file should now look like Figure 1.14.

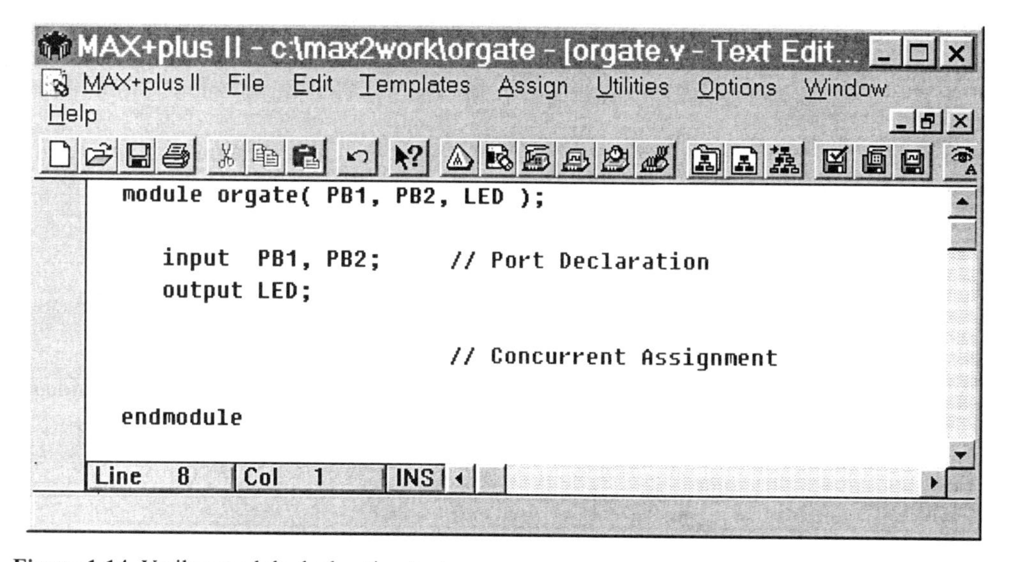

Figure 1.14 Verilog module declaration text.

Setting up the Behavioral Code

Click the mouse to just after the line starting with // Concurent Assignment. (We will be inserting another template here.) Following the earlier procedure for selecting a Verilog template (start with a right click), and select a

Continuous Assignment Statement. (A single assign statement will specify the internal logic of this design.) The syntax for a **Continuous Assignment Statement** appears in the text window after the other text. (You can now see why the template is left highlighted – had you not placed your cursor first, text would have appeared at your last cursor position. If you do misplace the template, hitting the delete key removes the highlighted text.)

Editing the Continuous Assignment Statement

Change the identify name in the assign statement to "LED" and value to:

! (! PB1 | ! PB2);

Verilog is based on C and "|" (vertical line) is the bit wise OR operator. The "!" (exclamation point) is the NOT operator. Delete the remaining comment lines that start with "//". Delete the ";" at the end of the assign LED statement (This causes a deliberate syntax error that will be detected and fixed later.) The file should now appear as in Figure 1.15.

Before You Compile

Before you compile the Verilog code, the device type and pin numbers need to be assigned. Select **Assign ⇨ Device**. Under **Device Family** select the **FLEX 10K** family. Under **Devices** select the **EPF10K20RC240-*X*** chip (-*X* is the speed grade of the chip. Most UP1 boards contain the –4 part).

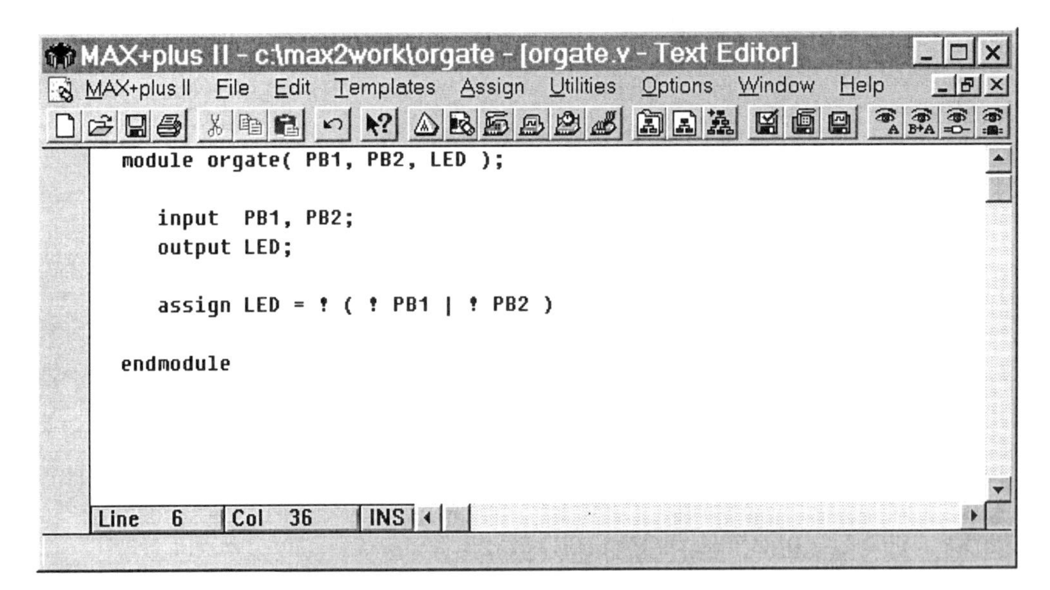

Figure 1.15 Verilog active low OR-gate model (with syntax error).

Next, select **Assign ⇨ Pin/Location/Chip**. Select **Pin** and scroll down to assign pin **28** to PB1. If your pins are already defined from the earlier tutorial just confirm the pin assignments. Repeat this process and assign or confirm that pin 29 is connected to PB2 and that pin 14 is connected to the decimal LED.

At this point, Verilog code is generally ready to be compiled, simulated, and downloaded to the board using steps identical to those used earlier in the schematic entry method. Once pin assignments are made, they are stored in the project's *.acf file.

1.8 Compiling the Verilog Design

The Compile process checks for syntax errors, synthesizes the logic design, produces timing information for simulation, fits the design on the CPLD, and generates the file required to program the CPLD. After any changes are made to the design files or pin assignments, the project should always be re-compiled prior to simulation or programming.

Select **File ⇨ Project ⇨ Set Project to Current File**. Next, compile by selecting **File ⇨ Project ⇨ Save and Compile**.

Checking for Compile Warnings and Errors

The project should compile with **1 Error**. After compiling the Verilog code, a window indicating an error should appear. (See Figure 1.16.)

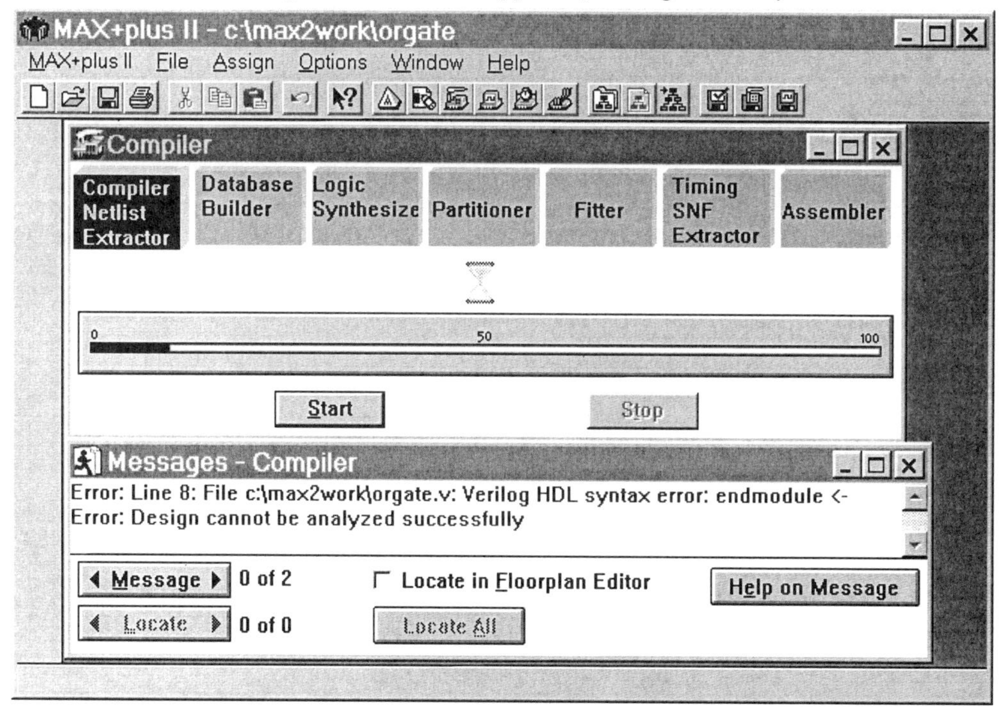

Figure 1.16 Verilog compilation with a syntax error.

Double click on the red error line and note that the cursor is placed in the editor either on or after the line missing the semicolon(;). Verilog statements should end with a semicolon. Add the semicolon to the end of the line so that it is now reads:

assign LED = ! (! PB1 | ! PB2);

Now, recompile, and you should have no errors.

1.9 Timing Analysis

With every physical device, there are timing considerations. A CPLD's timing is affected by:

- Input buffer delays,
- Routing interconnect delays within the CPLD,
- The internal logic delay (in this case the OR), and
- Output buffer delays.

The timing analysis tool can be used to determine:

- The physical delay times, and
- The maximum clock rates in your design.

Starting the Analyzer

At the upper left pull-down menu, select **MAX+PLUS II ⇨Timing Analyzer** and hit the **Start** button at the bottom of the **Timing Analyzer** window. A matrix of input to output delay times for the project will be computed and displayed as seen in Figure 1.17.

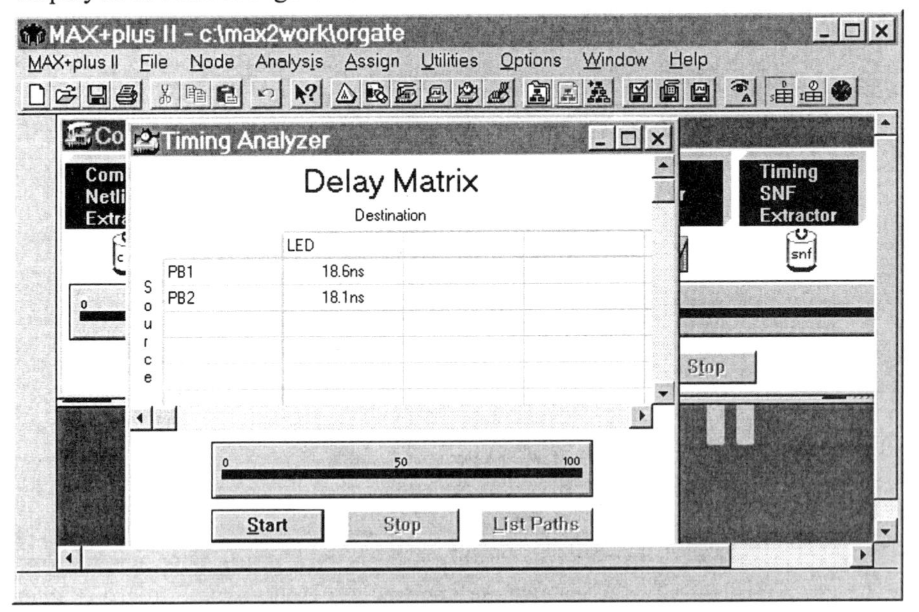

Figure 1.17 Timing analyzer showing input to output timing delays.

Note that this is the same delay time seen in the simulator. These times also include the input-to-output buffer delays and the interconnect delays inside the CPLD. The OR logic delay is just a few nanoseconds relative to the rest of the device delay. The actual time shown will vary with different versions of the Altera CAD tools and different chip speed grades. Other timing analysis options include setup times, hold times, and clock rates for sequential circuits.

1.10 The Floorplan Editor

The floorplan editor is a visual tool to assist expert users in manually placing and moving portions of logic circuits to different logic cells inside the CPLD. This is done in an attempt to achieve faster timing or better utilization of the CPLD. Floorplanning is typically used only on very large designs that contain subsections of hardware with critical high-speed timing.

For non-expert users, use of the compiler's automatic place-and-route tools is recommended. Automatic place-and-route was already performed in the compile process of the tutorial.

To see the compiler's automatic physical placement of the logic design inside the CPLD, select **MAX+PLUS II ⇨Floorplan Editor** and scroll around to find the colored pins and logic elements used in your design. Different versions of the software may place the logic in different locations inside the chip. One implementation of the OR-gate design is shown in Figure 1.18.

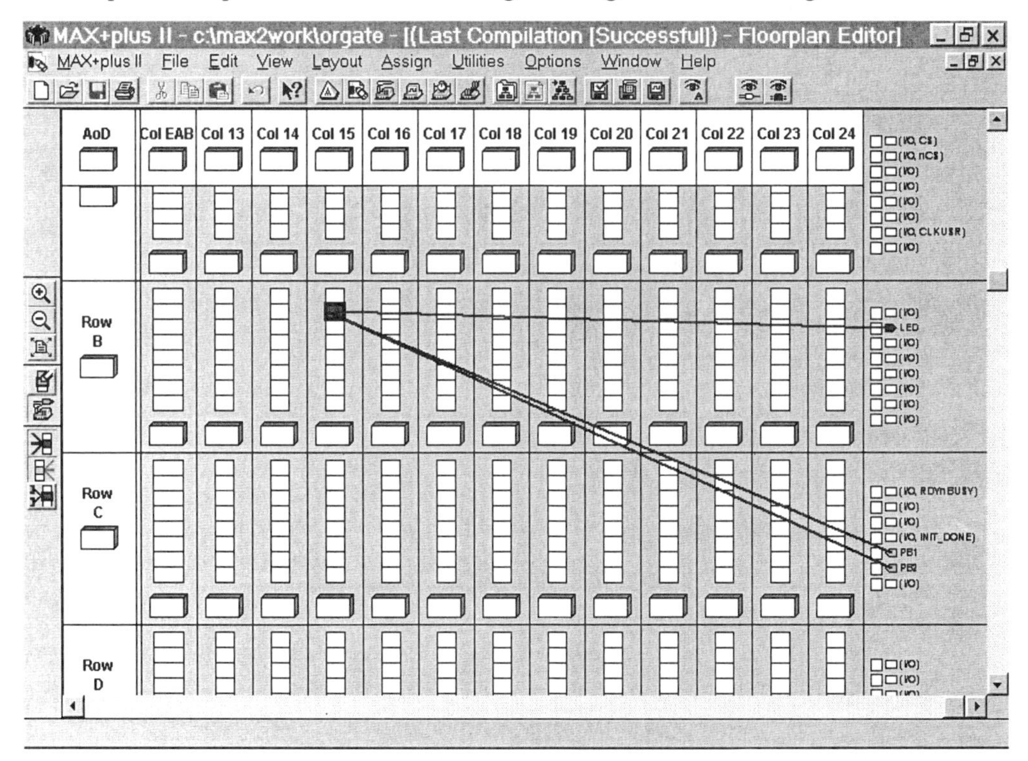

Figure 1.18 Floorplanner with internal CPLD placement of OR-gate logic cell and I/O pins.

There is a lot of empty space since the FLEX 10K20 can contain up to 20,000 logic gates. On a UP 1X board, the FLEX 10K70 can contain up to 70,000 gates. Pins and logic cells used in the design are color-coded. If you move the logic cell to another location, it will make small changes to the circuit timing because of changes in the interconnect delays inside the CPLD.

1.11 Symbols and Hierarchy

The symbol editor is used to edit or create a symbol to represent a logic circuit. Symbols are automatically created for a design whenever a VHDL or Verilog file is compiled.

Try **File ⇨ Open** select **orgate.sym** to see the new symbol for your VHDL based design as shown in Figure 1.19. Inputs are typically shown on the left side of the symbol and outputs on the right side. Symbols are used for design hierarchy in more complex schematics. This new symbol can be used to add the circuit to a design with the graphic editor just like the BNOR2 symbol used earlier in the tutorial. Clicking on a symbol in the graphic editor will take the user to the underlying logic circuit or HDL code that it represents.

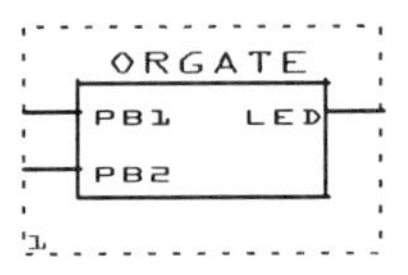

Figure 1.19 OR gate logic symbol.

1.12 Functional Simulation

In large designs with long compile and simulation times, another type of simulation is commonly used. A functional simulation does not include device delay times and it is used to check for logical errors only. A timing simulation that included device delay times was used earlier in the tutorial. After fixing logical errors with a functional simulation, a timing simulation is still necessary to check for any timing related errors in a design.

Performing a Functional Simulation

To perform functional simulation use **MAX+PLUS II ⇨Compiler** to open the compiler window. Then turn on **Processing ⇨Functional SNF extractor**. Then start the compiler and simulate. Open the SCF file and note that the output changes without delay in response to an input. To switch back to a timing simulation, unselect **Processing ⇨ Functional SNF extractor**, recompile, and simulate.

1.13 For additional information

This short tutorial has gone through the basics of a simple design using a common path through the design tools. As you continue to work with the tools, you will want to explore more of the menus, options and shortcuts. Chapter 4 will introduce a more complex design example. A MAX+PLUS II Getting Started reference manual with a more in depth tutorial is available free online at Altera's website, www.altera.com. A number of files are maintained in the project directory to support a design. Appendix B contains a list of different file extensions supported in MAX+PLUS II.

1.14 Laboratory Exercises

1. The tutorials ORed the active low signals from the pushbuttons and produced an output that was required to be low to turn on an LED. This was accomplished with the "negative-logic OR" gate illustrated to the left in Figure 1.20.

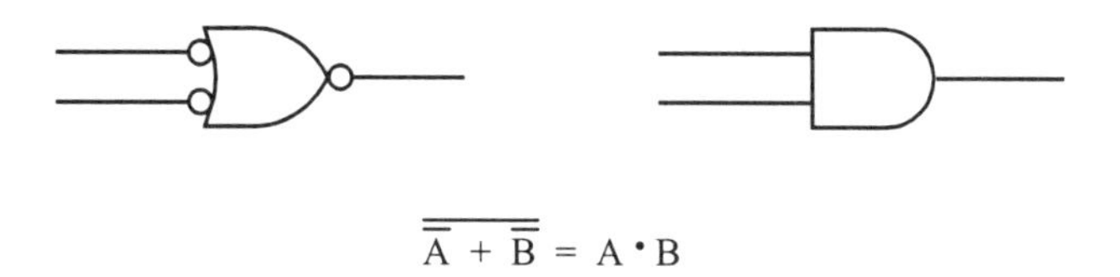

$$\overline{\overline{A} + \overline{B}} = A \cdot B$$

Figure 1.20 Equivalent gates: A negative logic OR and a positive-logic AND.

We know from DeMorgan's Law that the equation in Figure 1.20 represents an equivalence. We should therefore be able to substitute a simple two-input AND gate as illustrated in Figure 1.20 and accomplish the same task as the single gate used in the tutorial. Substitute the AND2 gate for the BNOR2 gate in the schematic capture, then compile, simulate, and download the AND circuit. What can you conclude?

2. Substitute in the VHDL code:

```
LED <= PB1 AND PB2;
```

Or into the Verilog code:

```
assign LED = PB1 & PB2;
```

Compile, simulate, and download the new circuit. What can you conclude about gate equivalence?

3. Switch the input waveforms on PB1 and PB2 in the waveform editor on any of the DeMorgan equivalent circuits. Note that one combination generates a short spike (glitch) while the other does not. Why does this happen?

4. Design a logic circuit to turn on the LED when both pushbuttons are pressed. Compile, simulate, and download the new circuit.

5. Try a different logic function such as XOR. Start at the beginning or edit your existing schematic by deleting and replacing the BNOR2 symbol. Next repeat the tutorial steps to compile, simulate, download and test.

6. Repeat problem 2 for all of the basic gates including, OR, NOR, NAND, XOR, XNOR, and NOT. Try using different parts of the seven-segment display and output your results simultaneously. Look up the pin connections to the FLEX chip in Appendix C and be sure to give each pinout a different name.

7. Design, enter, simulate and implement a more complex logic gate network. One suggestion is a half adder. You will need two outputs. The decimal point on the second LED display digit is connected to pin 25.

8. In the schematic editor, try building the design with some TTL parts from the maxplus2/max2lib/mf library. Search for help on "74" to find help files with data sheets on the TTL parts in the library.

9. Draw a schematic and develop a simulation to test the 2-to-1 Mux function in the mf library. Use MAX+PLUS II's Help function to see data sheets on mf library parts.

10. View the orgate.rpt file and find the device utilization, the pin assignments, and the netlist. A substantial portion of the time delay in this simple logic design is the input and output buffer delays and the internal routing of this signal inside the CPLD. Find this delay time by removing the BNOR2 gate and one of the inputs in the schematic. Connect the input pin to the output pin, recompile and rerun the timing analyzer to estimate this time delay.

11. Use the floorplan editor to move the logic cell used in the OR-gate design to another location inside the CPLD. Start the floorplan editor on the OR-gate project and use **LAYOUT ⇨ Current Assignments Floorplan** to edit the layout. The logic cell that performs the OR operation is moved to the unassigned node window at the top of the screen and it can now be dragged and dropped at another location inside the chip. Try moving it over two or three columns closer to the pushbutton and LED pins. Not all locations of the logic cell will work and some trial and error will be required. Recompile and rerun the timing analyzer and compare the resulting time delays with the original time delays. See if you are able to achieve a faster implementation than the automatic place-and-route tools.

12. Remove the pin number constraints from the schematic and let the compiler assign the pin locations. Rerun the timing analyzer and compare the time delays. Are they faster or slower than having specified the input pins?

13. Retarget the example design to the MAX chip. Pin numbers for the MAX decimal point LED can be found in Appendix C. It will be necessary to connect jumper wires from the MAX header to the pushbuttons. Select pins near the pushbuttons. Pin numbers can be seen on the board's silk-screen. Compare the timing from the MAX implementation to the FLEX implementation.

14. If a storage oscilloscope or a fast logic analyzer is available, compare the predicted delay times from the simulation and timing analysis to the actual delays measured on the UP 1 board. Use the MAX chip as the target device for this problem. Probes are easier to connect to the MAX device since it already has header sockets surrounding the chip. Jumper wires can be plugged into the header and the other end of the jumper wires can be connected to oscilloscope or logic analyzer probes.

15. Draw a schematic the uses the LPM_ADD_SUB megafunction to add two signed numbers on the FLEX device. In the **Enter Symbol** window, use the megawizard to help configure LPM symbols. Verify the proper operation using a simulation with two 4-bit numbers. Do not use pipelining, clock, or carry in. Vary the number of bits in the adder and find the maximum delay time using the timing analyzer. Plot delay time versus number of bits for several adder sizes. Using the LC percentages listed in the project's *.rpt file, estimate gate count by multiplying by 20,000. Plot gate count versus number of bits.

16. Use the DFF part from the prim library and enter the symbol in a schematic using the graphical editor. Develop a simulation that exercises all of the features of the D flip-flop. Use Help on DFF for more information on this primitive.

17. Use the DFFE part from the prim library and enter the symbol in a schematic using the graphical editor. Develop a simulation that exercises all of the features of the D flip-flop with a clock enable. Use Help on DFFE for more information on this primitive.

18. Use gates and a DFF part from the prim library with graphical entry to implement the state machine shown in the following state diagram. Verify correct operation with a simulation using the Altera CAD tools. The simulation should exercise all arcs in the state diagram. A and B are the two states, X is the output and Y is the input. Use the timing analyzer's **Analysis ⇨ Registered performance** option to determine the maximum clock frequency on the FLEX device. Reset is asynchronous and the DFF Q output should be high for state B.

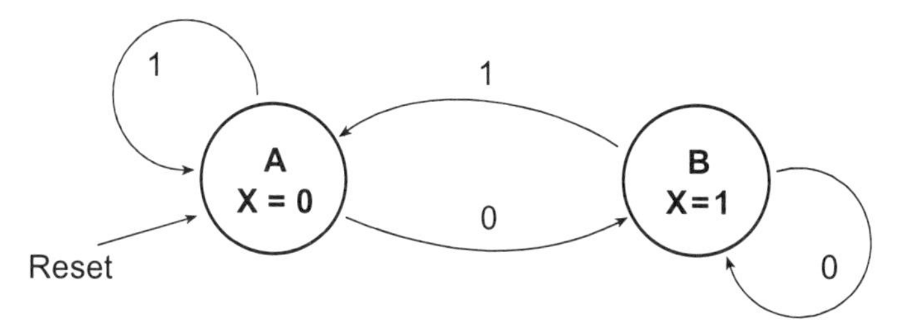

19. Repeat the previous problem but use one-hot encoding on the state machine. For one-hot encoding use two flip-flops with only one active for each state. For state A the flip-flop outputs would be "10" and for state B "01".

CHAPTER 2

The Altera UP 1 and UP 1X CPLD Boards

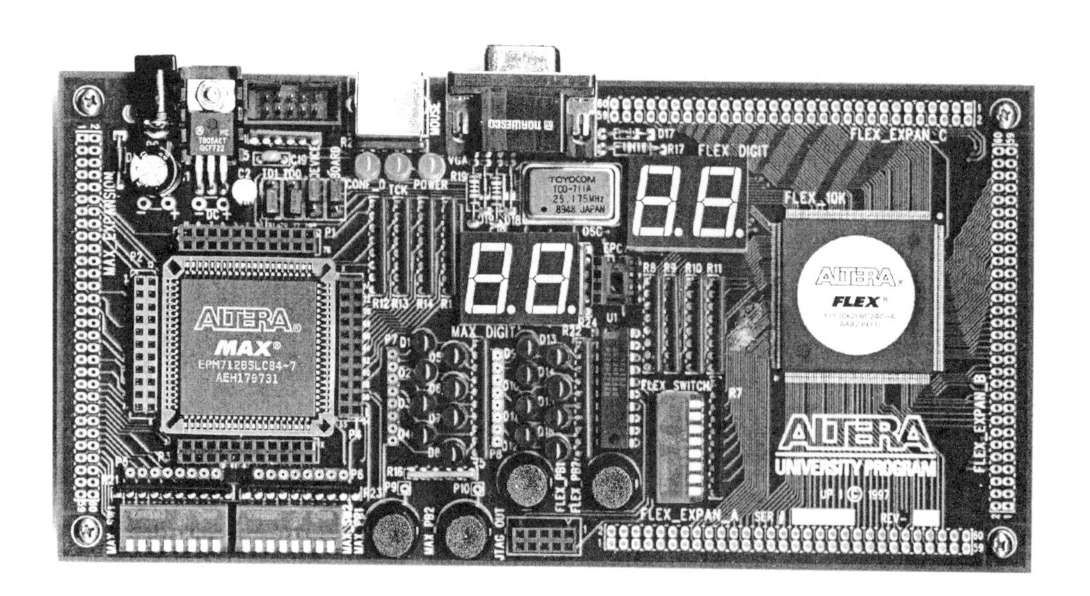

Photo: Altera UP 1 Board with two CPLDs

2 The Altera UP 1 and UP 1X CPLD Boards

The Altera University Program 1 (UP 1), CPLD design laboratory board is shown in Figures 2.1 and 2.2. This board supports both a MAX EPM7128S 2,500 gate, P-term based CPLD device, located on the left side, and a FLEX EPF10K20 20,000 gate, SRAM based CPLD device, located on the right side.

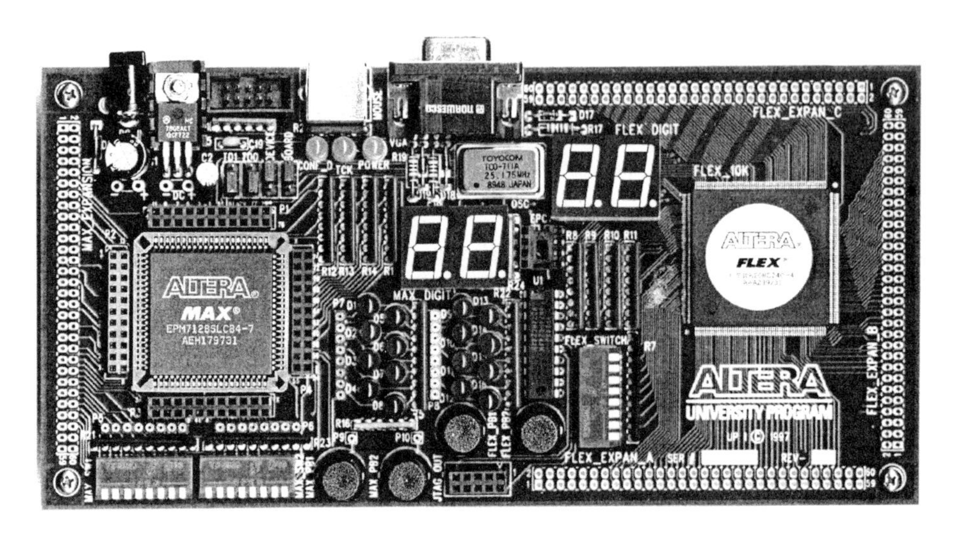

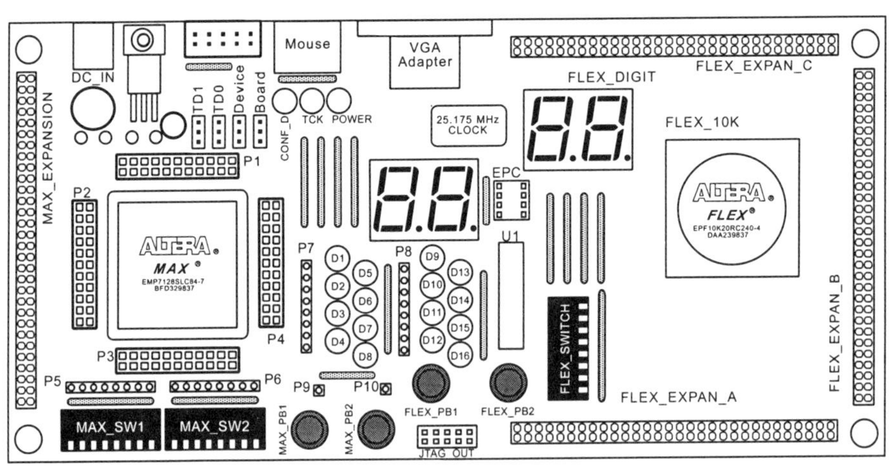

Figures 2.1 and 2.2 The Altera UP 1 and UP 1X board.

The newer UP 1X board is identical to the UP 1 except that it contains a larger FLEX 10K70 70,000 gate CPLD in place of the FLEX 10K20.

Either or both CPLD devices can be programmed using a JTAG ByteBlaster* cable attached to a PC printer port. Jumpers on the board select which device is programmed.

An external 7-9V DC power supply or a low-cost AC to DC wall adapter can be used to provide power. An on-board 25.175MHz clock oscillator is provided for use by the two CPLD devices and it is prewired to the MAX and FLEX chips at pins 83 and 91 respectively.

2.1 Programming Jumpers

A Byteblaster* programming adapter cable is supplied with the UP 1 board and it connects the UP 1 board to the PC's parallel port (LPT) for device programming. The printer port mode of the PC should be set in the PC's BIOS to ECP or EPP.

To select one of the CPLD devices to program, jumpers on the UP 1 board are set to select the MAX or the FLEX CPLD as shown in Table 2.1.

Table 2.1 UP 1 device selection jumpers for programming.

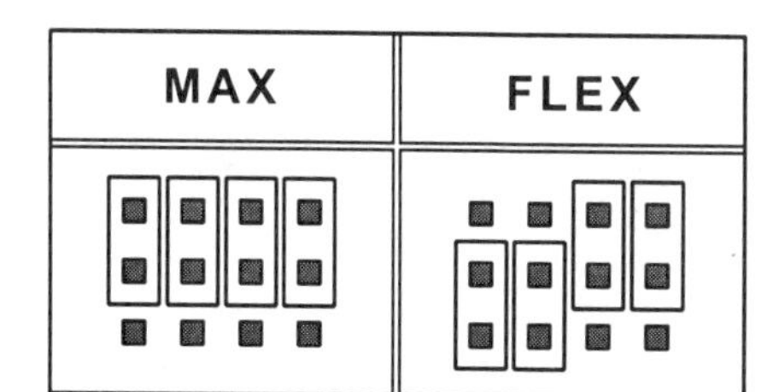

2.2 MAX 7000 Device and UP 1 I/O Features

The MAX EMP7128S 7000 device provides up to 2500 usable gates. It is on the left side of the UP 1 board and is in a socket. It connects to a prototyping header connector surrounding the chip. The UP 1's 25.175Mhz onboard clock is connected to a global clock buffer pin at pin 83. The MAX digit display has two seven-segment LED displays that are pre-wired to the device. Two eight-position DIP switches, sixteen discreet LEDs, and two SPST momentary contact non-debounced pushbuttons can also be connected to the MAX prototyping header with small wire jumpers. In addition, circuit-board holes are provided for an additional 60-pin expansion header that can be added to connect external hardware. Table 2.3 lists the MAX pin assignments for devices that are already connected to the MAX device.

2.3 MAX and FLEX Seven-segment LED Displays

The UP 1 board contains two dual seven-segment displays. One set is pre-wired to the MAX chip and the other is pre-wired to the FLEX chip. As seen in Figure 2.3, each display contains seven segments and one decimal point for a total of eight LEDs. Each LED is connected to and controlled by output pins on the MAX or FLEX chip. A low on the output pin turns on each LED.

The common connection of all on the display segments is already connected to 5 volts. This is a common practice that makes the LEDs brighter since TTL outputs traditionally provide more current at the low level.

By turning on various segments it is possible to display decimal and hexadecimal numbers in the display. For hexadecimal digits B and D, a lower case "b" and "d" is normally used to avoid confusion with "8" and "0".

Logic is needed to convert from 4-bit binary values to seven-segment values before displaying. The logic used to provide this output is called a seven-segment decoder and is provided in the UP1core libraries as dec_7seg. This circuit is used in Chapter 4.

This logic can also be designed at the gate level by putting the truth table in seven four-variable Karnaugh maps and minimizing the logic. As an alternative, it can be implemented by selecting a seven-segment decoder from the TTL library parts, or automatically minimized and synthesized using a language like VHDL, Verilog, or AHDL.

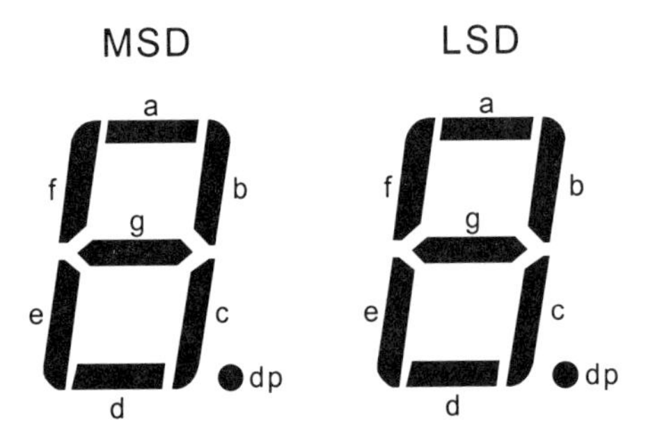

Figure 2.3 MAX and FLEX seven-segment LED display segment names.

The truth table for a decimal to seven-segment decoder is shown in Table 2.2.

Table 2.2 Truth table for seven-segment display.

Display	Bits 3-2-1-0	Seg a	Seg b	Seg c	Seg d	Seg e	Seg f	Seg g
0	0000	1	1	1	1	1	1	0
1	0001	0	1	1	0	0	0	0
2	0010	1	1	0	1	1	0	1
3	0011	1	1	1	1	0	0	1
4	0100	0	1	1	0	0	1	1
5	0101	1	0	1	1	0	1	1
6	0110	1	0	1	1	1	1	1
7	0111	1	1	1	0	0	0	0
8	1000	1	1	1	1	1	1	1
9	1001	1	1	1	1	0	1	1

Table 2.3 UP 1-board EPM7128S MAX CHIP I/O pin assignments.

Pin Name	Pin Type	Pin	Function of Pin
MSD_dp	OUTPUT PIN	68	Most Significant Digit of Seven-segment Display Decimal Point Segment (0 = LED ON, 1 = LED OFF)
MSD_g	OUTPUT PIN	67	MSD Display Segment G (0 = LED ON, 1 = LED OFF)
MSD_f	OUTPUT PIN	65	MSD Display Segment F (0 = LED ON, 1 = LED OFF)
MSD_e	OUTPUT PIN	64	MSD Display Segment E (0 = LED ON, 1 = LED OFF)
MSD_d	OUTPUT PIN	63	MSD Display Segment D (0 = LED ON, 1 = LED OFF)
MSD_c	OUTPUT PIN	61	MSD Display Segment C (0 = LED ON, 1 = LED OFF)
MSD_b	OUTPUT PIN	60	MSD Display Segment B (0 = LED ON, 1 = LED OFF)
MSD_a	OUTPUT PIN	58	MSD Display Segment A (0 = LED ON, 1 = LED OFF)
LSD_dp	OUTPUT PIN	79	Least Significant Digit of Seven-segment Display Decimal Point Segment (0 = LED ON, 1 = LED OFF)
LSD_g	OUTPUT PIN	77	LSD Display Segment G (0 = LED ON, 1 = LED OFF)
LSD_f	OUTPUT PIN	75	LSD Display Segment F (0 = LED ON, 1 = LED OFF)
LSD_e	OUTPUT PIN	76	LSD Display Segment E (0 = LED ON, 1 = LED OFF)
LSD_d	OUTPUT PIN	74	LSD Display Segment D (0 = LED ON, 1 = LED OFF)
LSD_c	OUTPUT PIN	73	LSD Display Segment C (0 = LED ON, 1 = LED OFF)
LSD_b	OUTPUT PIN	70	LSD Display Segment B (0 = LED ON, 1 = LED OFF)
LSD_a	OUTPUT PIN	69	LSD Display Segment A (0 = LED ON, 1 = LED OFF)
PB1	INPUT PIN	*	Push-Button 1 (non–debounced, 0 = button depressed)
PB2	INPUT PIN	*	Push-Button 2 (non–debounced, 0 = button depressed)
D1..D16 LEDs	OUTPUT PIN	*	16 Discrete LEDs - D1...D16 (0 = LED ON, 1 = LED OFF)
SW1 & SW2	INPUT PIN	*	MAX DIP Switch Inputs - SW*x*S1...SW*x*S8 (1 = Open, 0 = Closed)
Clock	INPUT PIN	83	25.175Mhz System Clock on low skew Global Clock Line
Prototyping Header Pins	INPUT, OUTPUT	1-84	Black Prototyping Headers next to MAX chip Numbers are silk-screened on board. Pins 12, 33, 54, 75, and 83 are not available.

* Jumper wires from the switch or LED to the MAX prototyping headers are required to use these devices.
Any available unused MAX header pin can be assigned to this device.

Note: A number of other pins are pre-assigned and required for programming, power, and ground connections.

WHEN CONNECTING EXTERNAL HARDWARE, ADDITIONAL PINS ARE AVAILABLE FOR USE ON THE MAX EXPANSION CONNECTOR ON THE BOARD. REFER TO THE UP 1 UNIVERSITY PROGRAM DESIGN LABORATORY PACKAGES USER GUIDE FOR DETAILS. THIS MANUAL IS SUPPLIED WITH THE UP 1 BOARD AND IS AVAILABLE FREE ALONG WITH A MANUAL SUPPLEMENT AT HTTP://WWW.ALTERA.COM/.

2.4 FLEX 10K Device and UP 1 I/O Features

The FLEX EPF10K20RC240 device provides up to 20,000 usable gates and is located on the right side of the UP 1 board. With six embedded array blocks (EABs), it can also provide up to 12K bits of memory. The larger FLEX EPF10K70 on the UP 1X board provides up to 70,000 usable gates and it contains nine EABs.

The UP 1's 25.175Mhz onboard clock is connected to a global clock buffer input at pin 91. This frequency has been selected to provide the exact frequency required to generate 640 by 480 VGA video. The clock can also be divided down using internal logic, when a slower clock frequency is needed.

The FLEX device is attached to a VGA video connector, a PS/2 mouse and keyboard port, two seven-segment LED displays, an eight-position DIP switch, and two SPST momentary contact non-debounced pushbuttons. The VGA and PS/2 connectors have no additional circuitry and provide simple pin connections to the FLEX device. To generate video output, mouse, or keyboard input, an interface must be designed using logic inside the FLEX device. The UP1core library functions that support VGA output along with keyboard and mouse inputs are described in Chapter 5. The internal design of these interface circuits is described in Chapters 9, 10, and 11.

In addition, circuit-board holes are provided for three 60-pin FLEX expansion headers that can be added to connect external hardware to the FLEX chip. One of these expansion headers is used for the robot described in Chapter 12.

The FLEX 10K20 is a surface mount chip and it is soldered directly to the board. It is difficult if not impossible to replace the FLEX chip, so extreme care should be exercised when interfacing FLEX I/O pins to any external devices. Table 2.4 and Appendix C lists the FLEX pin assignments for I/O devices that are already connected to the FLEX device.

WHEN CONNECTING EXTERNAL HARDWARE, ADDITIONAL PINS ARE AVAILABLE FOR USE ON THE FLEX EXPANSION A,B, AND C CONNECTORS ON THE BOARD. REFER TO THE UP 1, UNIVERSITY PROGRAM DESIGN LABORATORY PACKAGES USER GUIDE FOR DETAILS. THIS MANUAL COMES WITH THE UP 1 OR UP 1X BOARD AND IS AVAILABLE FREE ALONG WITH A MANUAL SUPPLEMENT AT HTTP://WWW.ALTERA.COM.

Table 2.4 UP 1 Board 10K20RC240 FLEX CHIP I/O pin assignments.

Pin Name	Pin Type	Pin	Function of Pin
MSD_dp	OUTPUT PIN	14	Most Significant Digit of Seven-segment Display - Decimal Point Segment (0 = LED ON, 1 = LED OFF)
MSD_g	OUTPUT PIN	13	MSD Display Segment G (0 = LED ON, 1 = LED OFF)
MSD_f	OUTPUT PIN	12	MSD Display Segment F (0 = LED ON, 1 = LED OFF)
MSD_e	OUTPUT PIN	11	MSD Display Segment E (0 = LED ON, 1 = LED OFF)
MSD_d	OUTPUT PIN	9	MSD Display Segment D (0 = LED ON, 1 = LED OFF)
MSD_c	OUTPUT PIN	8	MSD Display Segment C (0 = LED ON, 1 = LED OFF)
MSD_b	OUTPUT PIN	7	MSD Display Segment B (0 = LED ON, 1 = LED OFF)
MSD_a	OUTPUT PIN	6	MSD Display Segment A (0 = LED ON, 1 = LED OFF)
LSD_dp	OUTPUT PIN	25	Least Significant Digit of Seven-segment Display - Decimal Point Segment (0 = LED ON, 1 = LED OFF)
LSD_g	OUTPUT PIN	24	LSD Display Segment G (0 = LED ON, 1 = LED OFF)
LSD_f	OUTPUT PIN	23	LSD Display Segment F (0 = LED ON, 1 = LED OFF)
LSD_e	OUTPUT PIN	21	LSD Display Segment E (0 = LED ON, 1 = LED OFF)
LSD_d	OUTPUT PIN	20	LSD Display Segment D (0 = LED ON, 1 = LED OFF)
LSD_c	OUTPUT PIN	19	LSD Display Segment C (0 = LED ON, 1 = LED OFF)
LSD_b	OUTPUT PIN	18	LSD Display Segment B (0 = LED ON, 1 = LED OFF)
LSD_a	OUTPUT PIN	17	LSD Display Segment A (0 = LED ON, 1 = LED OFF)
FLEX_switch_1	INPUT PIN	41	FLEX DIP Switch Input 1 (1 = Open, 0 = Closed)
FLEX_switch_2	INPUT PIN	40	FLEX DIP Switch Input 2 (1 = Open, 0 = Closed)
FLEX_switch_3	INPUT PIN	39	FLEX DIP Switch Input 3 (1 = Open, 0 = Closed)
FLEX_switch_4	INPUT PIN	38	FLEX DIP Switch Input 4 (1 = Open, 0 = Closed)
FLEX_switch_5	INPUT PIN	36	FLEX DIP Switch Input 5 (1 = Open, 0 = Closed)
FLEX_switch_6	INPUT PIN	35	FLEX DIP Switch Input 6 (1 = Open, 0 = Closed)
FLEX_switch_7	INPUT PIN	34	FLEX DIP Switch Input 7 (1 = Open, 0 = Closed)
FLEX_switch_8	INPUT PIN	33	FLEX DIP Switch Input 8 (1 = Open, 0 = Closed)
PB1	INPUT PIN	28	Push-Button 1 (non–debounced, 0 = button depressed)
PB2	INPUT PIN	29	Push-Button 2 (non–debounced, 0 = button depressed)
Horiz_Sync	OUTPUT PIN	240	VGA Video Signal - Horizontal Synchronization
Vert_Sync	OUTPUT PIN	239	VGA Video Signal - Vertical Synchronization
Blue	OUTPUT PIN	238	VGA Video Signal - Blue Video Data
Green	OUTPUT PIN	237	VGA Video Signal - Green Video Data
Red	OUTPUT PIN	236	VGA Video Signal - Red Video Data
PS2_CLK	BIDIRECTIONAL	30	Clock line for PS/2 Mouse and Keyboard
PS2_DATA	BIDIRECTIONAL	31	Data line for PS/2 Mouse and Keyboard
Clock	INPUT PIN	91	25.175 MHz System Clock on low skew Global Clock Line

2.5 Obtaining a UP 1 or UP 1X Board and Power Supply

UP 1 and UP 1X boards are available for purchase from Altera at special student pricing. In addition to the board, there are some other miscellaneous parts that you will need to obtain elsewhere:

Power Supply

Jameco (http://www.jameco.com) product number, 127503 - DC910F5, is an AC to DC Wall Transformer that supplies 9VDC at 1000MA and has the proper power connector for the UP 1 and UP 1X boards.

Radio Shack also has some universal AC to DC Wall Transformers that will work, but be careful to get the correct voltage, current, polarity, and a female 2.5MM x 5.5 mm power connector that fits the UP 1 and UP 1X boards.

Supplies over 9V DC and up to 12V DC can be used but they tend to overheat the UP 1's 5V DC on board regulator. For proper voltage regulation, the UP 1's regulator requires a minimum input of 7 volts. If you plan to use a keyboard or mouse attached to the UP 1 or UP 1X board, a 1000 ma output is recommended, since the UP 1 or UP 1X must also provide power to them.

If you wish to use another power supply other than the one listed above and need a power connector to plug into the board, a coaxial DC power plug 5.5mm O.D. and 2.5mm ID is needed. For example, Radio Shack Number 274-1568C or equivalent will work. Be sure to double check the polarity when wiring up the connector. The outer conductor is ground and the inner conductor is +Vcc.

Heat Sink

The regulator chip gets very hot when large designs using several LEDs are running on the UP 1 or UP 1X. A small heat sink for the board's voltage regulator is highly recommended. DigiKey (http://www.digikey.com) part number, HS213, can be bolted to the regulator and is small enough to fit in the tight space so that the regulator can be bent back down on the board's surface.

Longer Cable

A longer 25pin to 25pin PC M/F parallel cable is also useful since the Byteblaster* cable provided with the UP 1 is often too short to reach the printer port. All 25 wires must be connected in the printer extension cable. Any computer store or electronic parts mail order catalog should have these cables.

PLCC Extraction Tool

Do not attempt to remove the MAX chip from the socket without a PLCC extraction tool. PLCC extraction tools are available from Digikey. The AMP 84 pin PLCC extraction tool works well for this purpose. This tool is only needed for replacing an inoperative MAX CPLD. One PLCC extraction tool per laboratory is generally sufficient.

CHAPTER 3

Programmable Logic Technology

Photo: An Altera FLEX 10K100 CPLD containing 10,000,000 Transistors and 100,000 gates. The CPLD is in a pin grid array (PGA) package. The cover has been removed so that the chip die is visible in the center of the package.

3 Programmable Logic Technology

A wide spectrum of devices is available for the implementation of digital logic designs as shown in Figure 3.1. Traditional off-the-shelf integrated circuit chips, such as SSI and MSI TTL, perform a fixed operation defined by the device manufacturer. A user must connect different chip types to build a circuit. Application specific integrated circuits (ASICs), complex programmable logic devices (CPLDs), and field programmable gate arrays (FPGAs) are integrated circuits whose internal functional operation is defined by the user. ASICs require a final customized manufacturing step for the user-defined function. A CPLD or FPGA requires user programming to perform the desired operation.

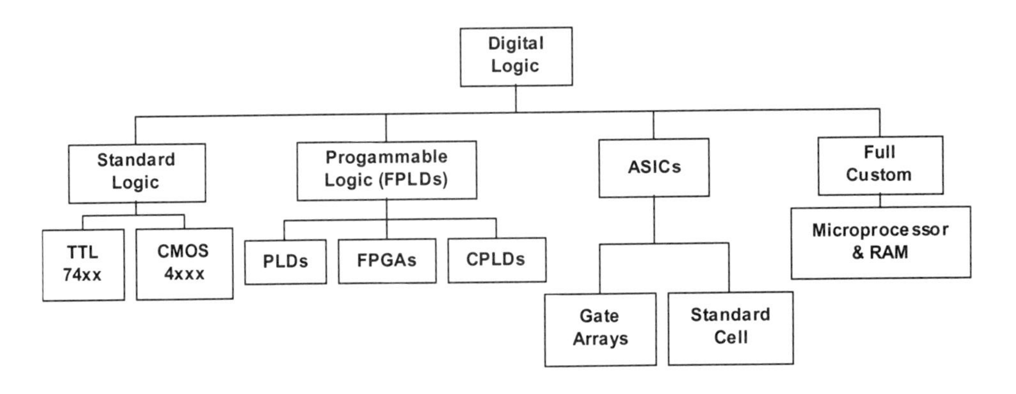

Figure 3.1 Digital logic technologies.

The design tradeoffs of the different technologies are seen in Figure 3.2. Full custom VLSI development of a design at the transistor level can require several years of engineering effort for design and testing. Such an expensive development effort is warranted only for the highest volume devices. This approach can generate the highest performance devices. Examples of full custom devices include the microprocessor and RAM chips used in PCs.

ASICs can be divided into two categories, Gate Arrays and Standard Cells. Gate Arrays are built from arrays of pre-manufactured logic cells. A single logic cell can implement a few gates or a flip-flop. A final manufacturing step is required to interconnect the sea of logic cells on a gate array. This interconnection pattern is created by the user to implement a particular design. Standard Cell devices contain no fixed internal structure. For standard cell devices, the manufacturer creates a custom photographic mask to build the chip based on the user's selection of devices, such as controllers, ALUs, RAM, ROM, and microprocessors from the manufacturer's standard cell library.

Since ASICs require custom manufacturing, additional time and development costs are involved. Several months are normally required and substantial setup fees are charged. Additional effort in testing must be performed by the user since chips are tested after the final custom-manufacturing step. Any design error in the chip will lead to additional manufacturing delays and costs. For products with long lifetimes and large volumes, this approach has a lower cost per unit than CPLDs or FPGAs. Economic and performance tradeoffs between ASICs, CPLDs, and FPGAs are constantly changing with each new generation of devices and design tools.

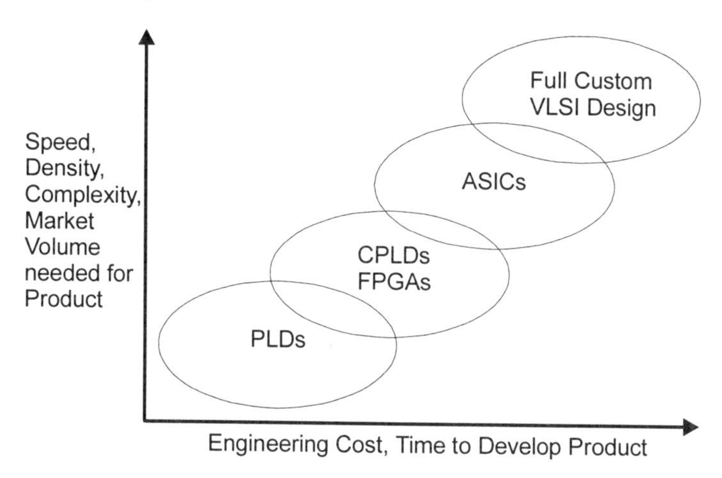

Figure 3.2 Digital logic technology tradeoffs.

Simple programmable logic devices (PLDs), such as programmable array logic (PALs), and programmable logic arrays (PLAs), have been in use for over twenty years. An example of a small PLA is shown in Figure 3.3. First, the logic equation is minimized and placed in sum of products (SOP) form. The PLA has four inputs, A, B, C, and D shown in the upper left corner of Figure 3.3. Every input connects to an inverter, making the inverted values of A, B, C, and D available for use. Each product term is implemented using an AND gate with several inputs. Outputs from the two product term's AND gates then feed into an OR gate.

A special shorthand notation is used in PLAs and PALs to represent the large number of inputs present in the AND and OR gate arrays. A gate input is present at each point where the vertical and horizontal signal lines cross in Figure 3.3. Note that this means that the two AND gates actually have eight inputs and the OR gate has two inputs in the PLA. Every input signal and it's complement is available as an input to the AND gates. Each gate input in the PLA is controlled by a fuse. Initially all fuses are intact. By blowing selected fuses, or programming the PLA, the desired SOP equation is produced. The top AND gate in Figure 3.3 has fuses intact to the A and B inputs, so it produces the AB product term. The lower AND gate has fuses set to produce $C\overline{D}$. The OR gate has both fuses intact, so it ORs both product terms from the AND gates to produce the final output, $F = AB + C\overline{D}$.

Small PLDs can replace several older fixed function TTL-style parts in a design. Most PLDs contain a PLA-like structure in which a series of AND gates with selectable or programmable inputs, feed into an OR gate. In PALs, the OR gate has a fixed number of inputs and is not programmable. The AND gates and OR gate are programmed to directly implement a sum-of-products Boolean equation. On many PLDs, the output of the OR gate is connected to a flip-flop whose output can then be feed back as an input into the AND gate array. This provides PLDs with the capability to implement simple state machines. A PLD can contain several of these AND/OR networks.

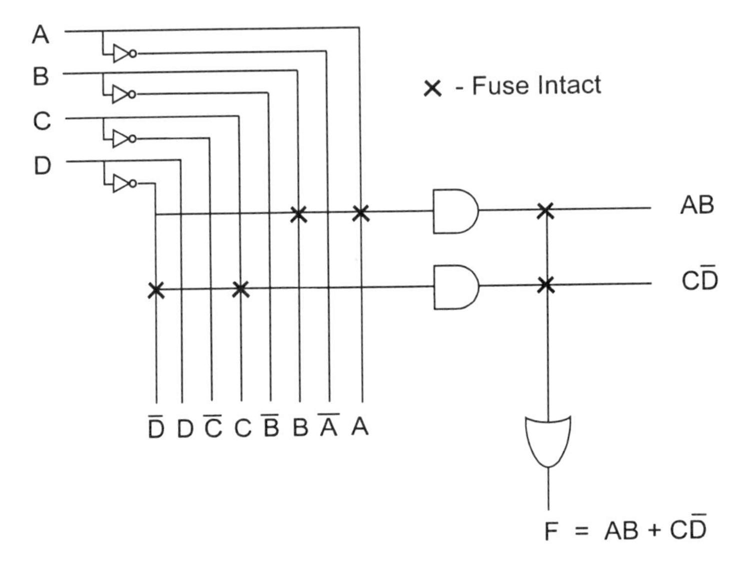

Figure 3.3 Using a PLA to implement a Sum of Products equation.

In more recent times, higher densities, higher speed, and cost advantages have enabled the use of programmable logic devices in a wider variety of designs. CPLDs and FPGAs are the highest density and most advanced programmable logic devices. Designs using a CPLD or FPGA typically require several weeks of engineering effort instead of months. These devices are collectively called field programmable logic devices (FPLDs).

ASICs and full custom designs provide faster clock times than CPLDs or FPGAs since they are hardwired and do not have programmable interconnect delays. Since ASICs and full custom designs do not require programmable interconnect circuitry they use less chip area and have a lower per unit manufacturing cost in large volumes. Initial engineering costs for ASICs and full custom designs are higher.

For all but the most time critical design applications, CPLDs and FPGAs have adequate speed with maximum clock rates typically in the range of 50-400Mhz; however, clock rates up to 1Ghz have been achieved on new generation FPLDs.

3.1 CPLDs and FPGAs

Internally, CPLDs and FPGAs typically contain multiple copies of a basic programmable logic element (LE) or cell. The logic element can implement a network of several logic gates that then feed into 1 or 2 flip-flops. Logic elements are arranged in a column or matrix on the chip. To perform more complex operations, logic elements can be automatically connected to other logic elements on the chip using a programmable interconnection network. The interconnection network is also contained in the CPLD or FPGA.

The interconnection network used to connect the logic elements contains row and/or column chip-wide interconnects. In addition, the interconnection network often contains shorter and faster programmable interconnects limited only to neighboring logic elements.

When a design approaches the device size limits, it is possible to run out of either gate, interconnect, or pin resources when using a CPLD or FPGA. CPLDs tend to have faster and more predictable timing properties while FPGAs offer the highest gate densities.

Clock signals in large FPLDs normally use special low-skew global clock buffer lines. This is a dedicated pin connected to an internal high-speed bus. This special bus is used to distribute the clock signal to all flip-flops in the device at the same time to minimize clock skew. If the global clock buffer line is not used, the clock is routed through the chip just like a normal signal. The clock signal could arrive at flip-flops at widely different times since interconnect delays will vary in different parts of the chip. This delay time can violate flip-flop setup and hold times and can cause metastability or unpredictable operation in flip-flops. Many large designs with a common clock that is used throughout the FPLD will require the use of the global clock buffer.

The size of CPLDs and FPGAs is typically described in terms of useable or equivalent gates. This refers to the maximum number of two input NAND gates available in the device. This should be viewed as a rough estimate of size only. Actual gate utilization of a particular design can vary significantly.

Figure 3.4 Examples of FPLDs and advanced high pin count package types.

The internal architecture of three examples of CPLD and FPGA device technologies, the Altera MAX 7000, the Altera FLEX 10K, and the Xilinx 4000 family will now be examined. An example of each of these devices is shown in Figure 3.4. From left to right the chips are an Altera MAX 7128S CPLD in a Plastic J-Lead Chip Carrier (PLCC), an Altera FLEX 10K70 CPLD in a Plastic Quad Flat Pack (PQFP), and a Xilinx XC4052 FPGA in a ceramic Pin Grid Array Package (PGA). The PGA package has pins on .1" centers while the PQFP has pins on .05" centers at the edges of the package. Both Altera and Xilinx devices are available in a variety of packages.

Packaging can represent a significant portion of the FPLD chip cost. The number of I/O pins on the FPLD package often limits designs. Larger ceramic packages such as a PGA with more pins are more expensive than plastic.

3.2 Altera MAX 7000S Architecture – A Product Term CPLD Device

The multiple array matrix (MAX) 7000S is a CPLD device family with 600 to 20,000 gates. This device is configured by programming an internal electrically erasable programmable read only memory (EEPROM). Since an EEPROM is used for programming, the configuration is retained when power is removed. This device also allows in-circuit reprogrammability.

The 7000 device family contains from 32 to 256 macrocells of the type seen in Figure 3.5. Similar to the early PALs, an individual macrocell contains five programmable AND gates with wide inputs that feed into an OR gate with a programmable inversion at the output. Just like a PAL, the AND/OR network is designed to implement Boolean equations expressed in sum-of-products form. Inputs to the wide AND gate are available in both normal and inverted forms.

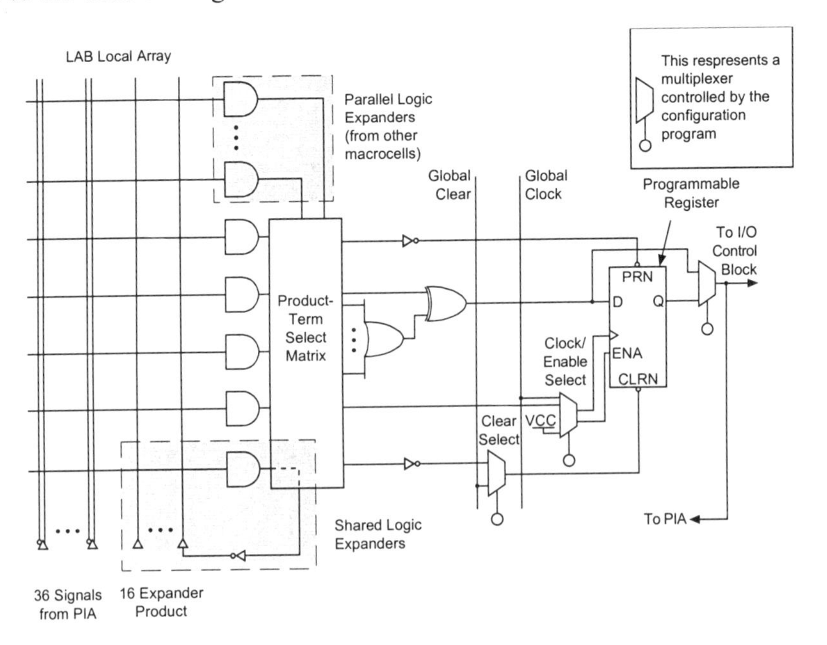

Figure 3.5 MAX 7000 macrocell.

Parallel expanders are included that are used to borrow extra product terms from adjacent macrocells for logic functions needing more than five product terms.

The output from the AND/OR network can then be fed into a programmable flip-flop. Inputs to the AND gates include product terms from other macrocells in the same local block or signals from the chip-wide programmable interconnect array (PIA). The flip-flop contains Bypass, Enable, Clear and Preset functions and can be programmed to act as a D flip-flop, Toggle flip-flop, JK flip-flop, or SR latch.

Macrocells are combined into groups of 16 and called logic array blocks (LABs), for the overall device architecture as shown in Figure 3.6. The PIA can be used to route data to or from other LABs or external pins on the device. Each I/O pin contains a programmable tri-state output buffer. A CPLD's I/O pin can thus be programmed as input, output, output with a tri-state driver, or even tri-state bi-directional.

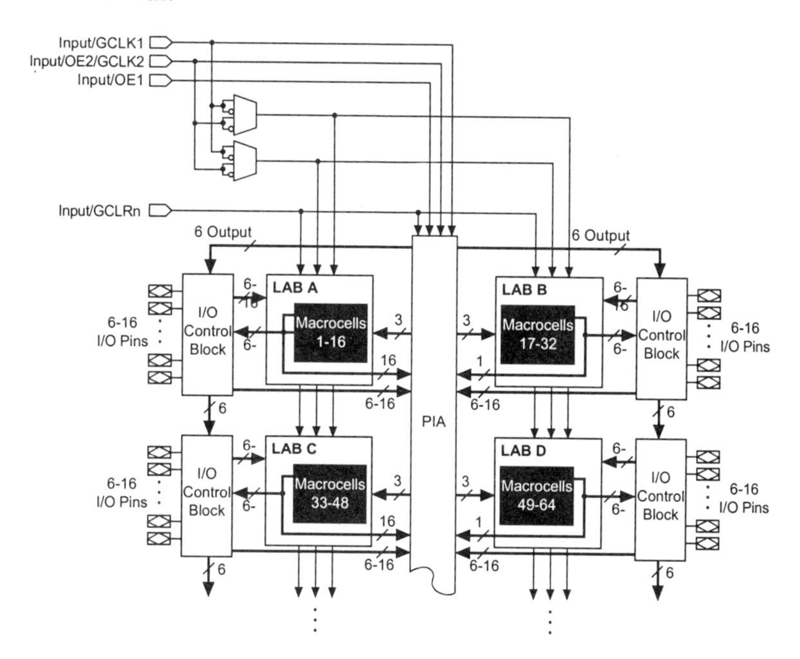

Figure 3.6 MAX 7000 CPLD architecture.

3.3 Altera FLEX 10K Architecture – A Look-Up Table CPLD Device

The flexible logic element matrix (FLEX), is a CPLD device family with 10,000 to 250,000 gates. Row and column interconnect lines can be seen in a photo of the chip die as shown in Figure 3.7. The FLEX device is configured by loading internal static random access memory (SRAM). Since SRAM is used, the configuration will be lost whenever power is removed. In actual systems, a small external low-cost serial programmable read only memory (PROM) is normally used to automatically load programming information when the device powers up.

Figure 3.7 FLEX 10K100 FPLD die photo, PIA interconnects are visible.

Figure 3.8 shows a FLEX logic element. Gate logic is implemented using a lookup table (LUT). The LUT is a high-speed 16 by 1 SRAM. Four inputs are used to address the LUT's memory. The truth table for the desired gate network is loaded into the LUT's SRAM during programming. A single LUT can therefore model any network of gates with four inputs and one output.

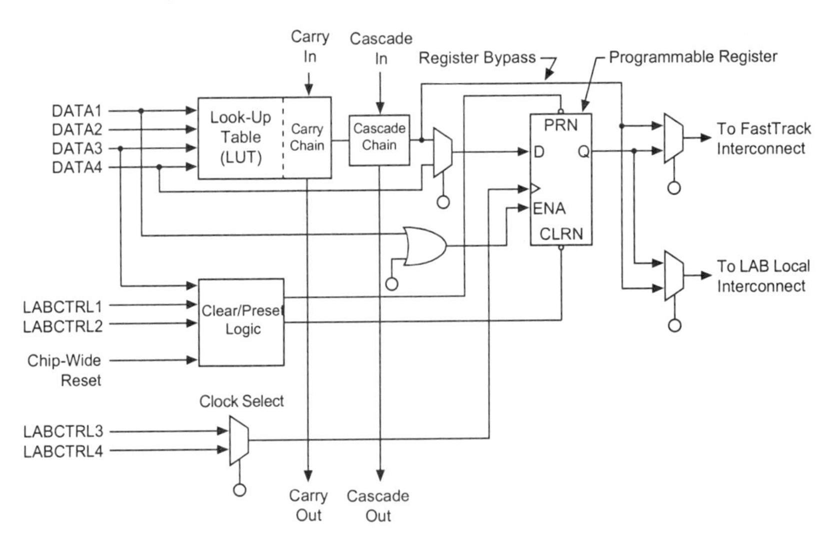

Figure 3.8 FLEX 10K Logic Element (LE).

An example showing how a LUT can model a gate network is shown in Figure 3.9. First, the gate network is converted into a truth table. Since there are four inputs and one output, a truth table with 16 rows and one output is needed. The truth table is then loaded into the LUT's 16 by 1 high-speed SRAM when the FPLD is programmed. Note that the four gate inputs, A, B, C, and D, are used as address lines for the RAM and that F, the output of the truth table, is the data that is stored in the LUT's RAM. In this manner, the LUT's RAM implements the gate network by performing a RAM based table lookup instead of using actual logic gates.

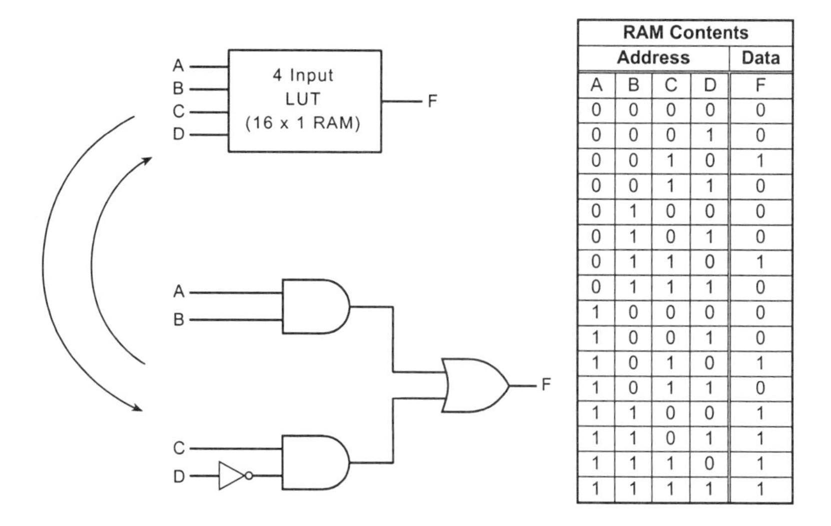

RAM Contents				
Address				Data
A	B	C	D	F
0	0	0	0	0
0	0	0	1	0
0	0	1	0	1
0	0	1	1	0
0	1	0	0	0
0	1	0	1	0
0	1	1	0	1
0	1	1	1	0
1	0	0	0	0
1	0	0	1	0
1	0	1	0	1
1	0	1	1	0
1	1	0	0	1
1	1	0	1	1
1	1	1	0	1
1	1	1	1	1

Figure 3.9 Using a lookup table (LUT) to model a gate network.

More complex gate networks require interconnections with additional neighboring logic elements. The output of the LUT can be fed into a D flip-flop and then to the interconnection network. The clock, Clear, and Preset can be driven by internal logic or an external I/O pin. The flip-flop can be programmed to act as a D flip-flop, T flip-flop, JK flip-flop, or SR latch. Carry and Cascade chains connect to all LEs in the same row.

Figure 3.10 shows a logic array block (LAB). A logic array block is composed of eight logic elements (LEs). Both programmable local LAB and chip-wide row and column interconnects are available. Carry chains are also provided to support faster addition operations.

Figure 3.11 shows the FLEX 10K device architecture. A matrix of LABs, and embedded array blocks (EABs), are connected via programmable row and column interconnects. Devices in the FLEX 10K family contain from 72 to 1,520 LABs.

An EAB contains 2048 bits of memory. FLEX devices contain from three to twenty EABs. Each EAB can be configured as 256 by 8, 512 by 4, 1024 by 2, or 2048 by 1 SRAM or ROM. In some cases, EABs can be used to implement gate level logic. As an example, a 4 by 4 multiplier can be built by storing the multiply truth table in a single EAB.

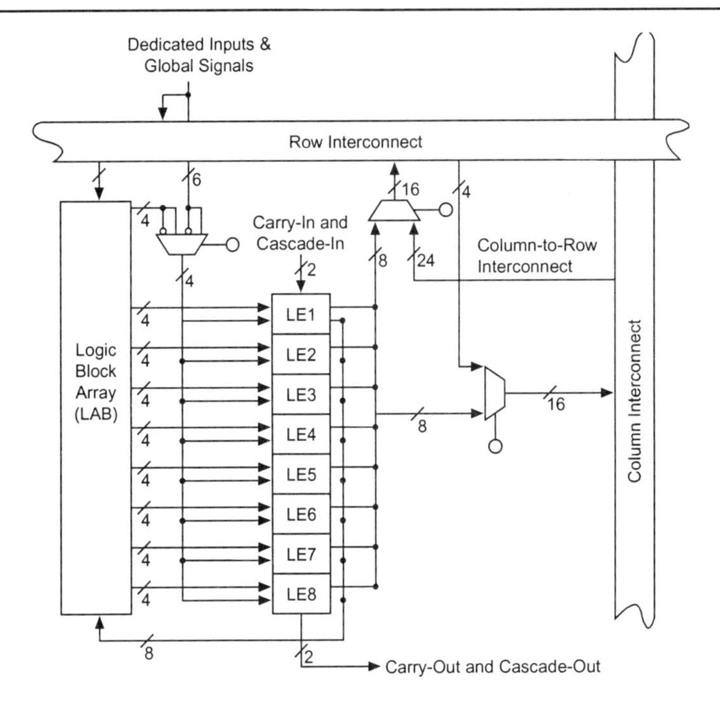

Figure 3.10 FLEX 10K Logic Array Block (LAB).

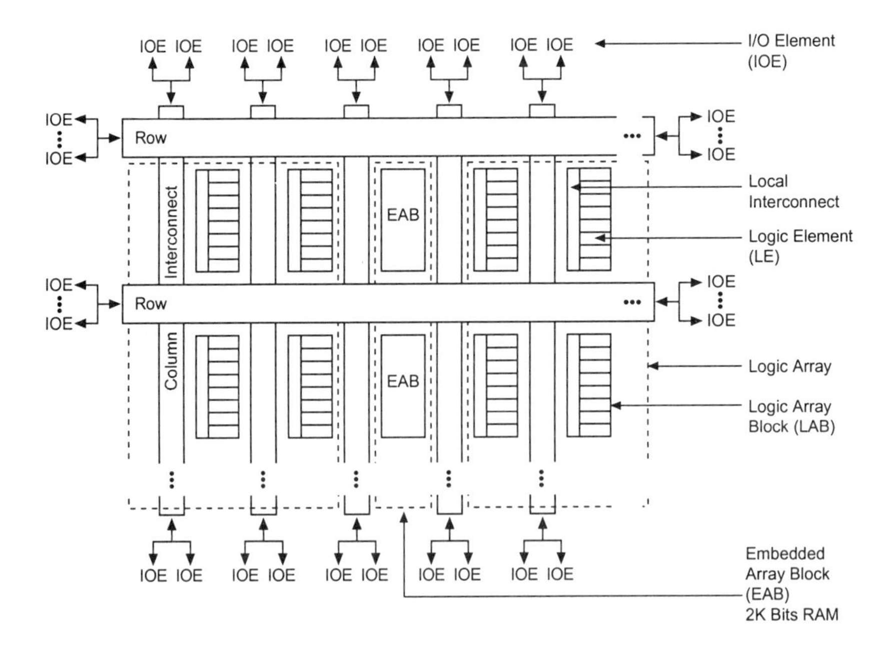

Figure 3.11 FLEX 10K CPLD architecture.

Input-output elements (IOEs) are located at each of the device's I/O pins. IOEs contain a programmable tri-state driver and an optional 1-bit flip-flop register. Each I/O pin can be programmed as input, output, output with a tri-state driver, or even tri-state bi-directional with or without a register. Two low-skew global clock buffer lines are also provided on the device.

3.4 Xilinx 4000 Architecture – A Look-Up Table FPGA Device

The Xilinx 4000 Family is an FPGA device family with 2,000 to 180,000 usable gates. It is configured by programming internal SRAM. Figure 3.12 is a photograph of a six-inch silicon wafer containing several XC4010E 10,000 gate FPGA chip dice. Figure 3.13 is a contrast-enhanced view of a single XC4010E die. If you look closely, you can see the 20 by 20 array of logic elements and the surrounding interconnect lines. Die that pass wafer-level inspection and testing are sliced from the wafer and packaged in a chip. FPLD yields are typically 90% or higher after the first few production runs.

As seen in Figure 3.14, this device contains a more complex logic element called a configurable logic block (CLB). Each CLB contains three SRAM-based lookup tables. Outputs from the LUTs can be fed into two flip-flops and routed to other CLBs. A CLB's lookup tables can also be configured to act as a 16 by 2 RAM or a dual-port 16 by 1 RAM. High-speed carry logic is provided between adjacent CLBs.

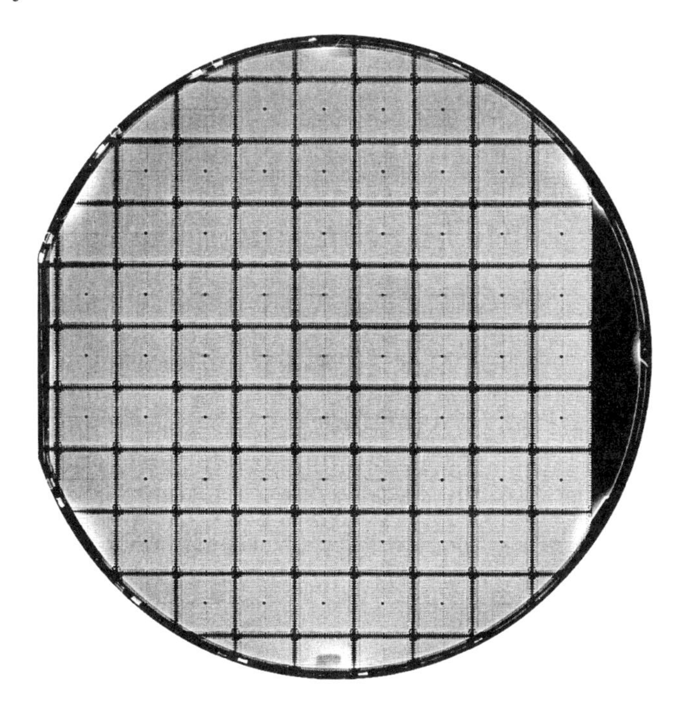

Figure 3.12 Silicon wafer containing XC4010E 10,000 gate FPGAs.

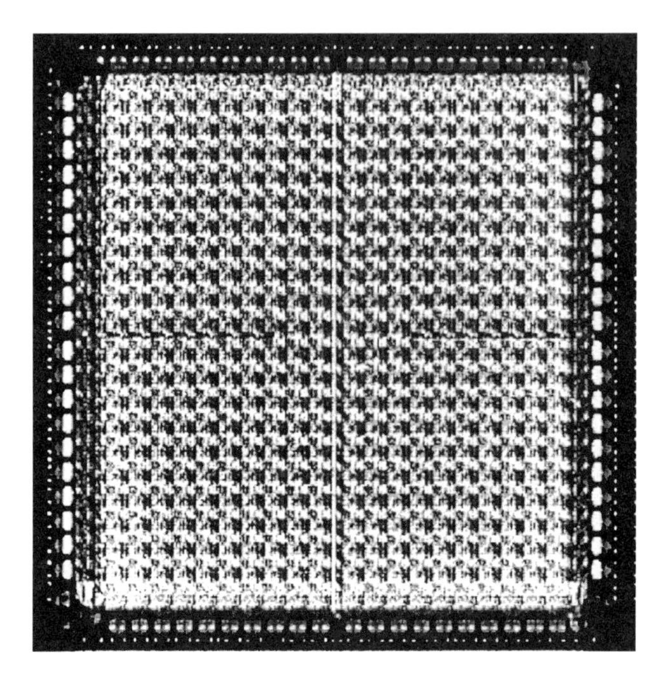

Figure 3.13 Single XC4010E FPGA die showing 20 by 20 array of logic elements and interconnect.

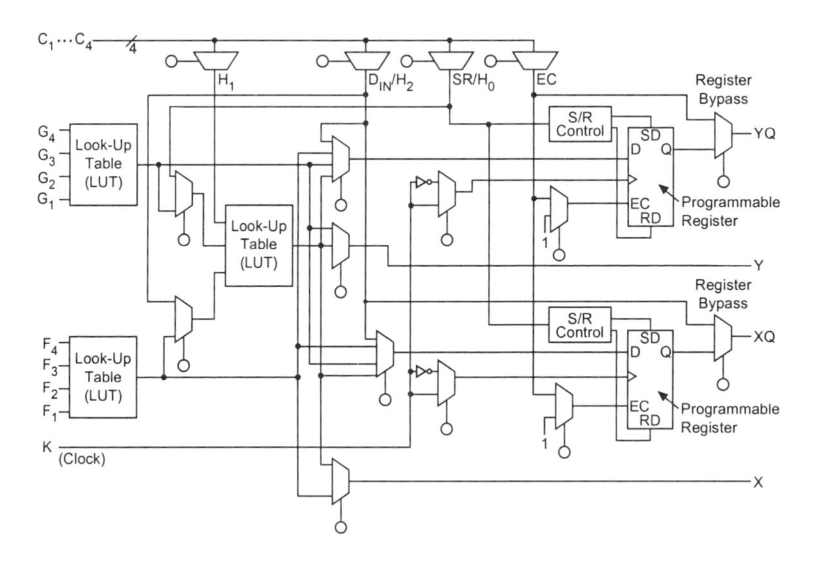

Figure 3.14 Xilinx 4000 Family Configurable Logic Block (CLB).

CLBs are arranged in a square matrix with a programmable hierarchical interconnection network. Devices in the family contain from 100 to 3,136 CLBs. The complex hierarchical interconnection network contains varying length row, column, and neighboring CLB interconnect structures. Eight low-skew global clock buffers are also provided. Input-output blocks (IOBs), contain programmable tri-state drivers and optional registers. Each I/O pin can be programmed as input, output, output with a tri-state driver, or tri-state bi-directional with or without a register.

3.5 Computer Aided Design Tools for Programmable Logic

Increasing design complexity and higher gate densities are forcing digital designs to undergo a paradigm shift. Old technology, low-density logic families, such as the TTL 7400 series are rarely used in new designs. With logic capacities of an individual FPLD chip approaching 10,000,000 gates, manual design at the gate level is no longer a viable option in complex systems. Rapid prototyping using hardware description languages (HDLs) and logic synthesis tools are quickly replacing traditional gate-level design with schematic capture entry. These new HDL-based logic synthesis tools can be used for ASIC, CPLD, and FPGA-based designs. The two most widely used HDLs at the present time are VHDL and Verilog.

The typical FPLD CAD tool design flow is shown in Figure 3.15. After design entry using an HDL or schematic, the design is automatically translated, optimized, synthesized, and saved as a netlist. (A netlist is a text-based representation of a logic diagram.) A functional simulation step is often added prior to the synthesis step to speed up simulations of large designs.

An automatic tool then fits the design onto the device by converting the design to use the FPLD's logic elements, first by placing the design in specific logic element locations in the FPLD and then by selecting the interconnection network routing paths. The place and route process can be quite involved and can take several minutes to compute on large designs. On large devices, combinatorial explosion (exponential growth) will prevent the tool from examining all possible place and route combinations. When designs require critical timing, some tools support timing constraints that can be placed on critical signal lines. These optional constraints are added to aid the place and route tool in finding a design placement with improved performance.

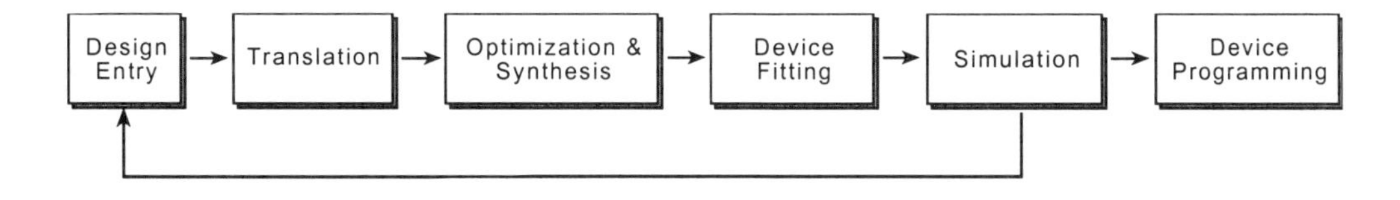

Figure 3.15 CAD tool design flow for FPLDs.

After place and route, simulation can be performed using actual gate and interconnect time delays from a detailed timing model of the device. Although errors can occur at any step, the most common path is to find errors during an exhaustive simulation. The final step is device programming and hardware verification on the FPLD.

3.6 Next Generation FPLD CAD tools

A few HDL synthesis tools now support behavioral synthesis. Unlike the more widely used register transfer level (RTL) models contained in this book, behavioral synthesis models do not specify the exact states and sequence of register transfers. A separate constraint file specifies the number of clocks needed to obtain selected signals and the tool automatically generates the state machines and register transfers needed.

Although not currently in widespread use, newer FPLD CAD tools are also appearing based on other languages such as C and Java. Some of these system-level tools output VHDL or Verilog models as an intermediate step. Tools that automatically generate an FPLD design from other engineering tools such as MATLAB are also being introduced.

3.7 Applications of FPLDs

The last decade has seen ever increasing application areas for FPLDs. A recent market study found over ten times as many new FPLD-based designs as ASIC-based designs. New generation FPLDs can have nearly ten million gates with clock rates approaching 1GHz. Example application areas include single chip replacements for old multichip technology designs, Digital Signal Processing (DSP), image processing, multimedia applications, high-speed communications and networking equipment such as routers and switches, the implementation of bus protocols such as peripheral component interconnect (PCI), microprocessor glue logic, co-processors, and microperipheral controllers.

Several large FPLDs with an interconnection network are used to build hardware emulators. Hardware emulators are specially designed commercial devices used to prototype and test complex hardware designs that will later be implemented on gate arrays or custom devices. Hardware emulators are commonly used to build a prototype quickly during the development and testing of microprocessors. Several of the recent Intel and AMD processors used in PCs were tested on hardware emulators before the full custom VLSI processor chip was produced.

A newer application area is reconfigurable computing. In reconfigurable computing, FPLDs are quickly reprogrammed or reconfigured multiple times during operation to enable them to perform different computations at different times for a particular application.

3.8 Features of New Generation FPLDs

Each new generation of FPLDs increases in size and performance. In addition to more logic elements, embedded memory blocks, and interconnect other new features are appearing. Some FPLDs contain a mix of both product term and lookup tables to implement logic. Such product term structures typically require less chip area to implement the complex gating logic present in large state machines and address decoders. Many FPLDs include several phase-locked loops (PLLs). These PLLs are used to multiply and adjust high-speed clock signals. Similar to microprocessors used in PCs, many new FPLDs use a lower 1.5 to 3 Volt internal core power supply. To easily interface to external processor and memory chips, new FPLDs feature selectable I/O standards on I/O pins.

High-speed hardware multipliers are also available in FPLD families targeted for multiply intensive DSP and graphics applications. Several FPLDs from Altera and Xilinx are available with commercial internal RISC microprocessor intellectual property (IP) cores. These include the NIOS, ARM, MIPS, and PowerPC. The NIOS processor is an HDL model that is synthesized using the FPLDs standard logic elements. The ARM, MIPS, and PowerPC are commercial IP cores with custom VLSI layouts. Several processors can be implemented in a single FPLD. These FPLDs come with additional software tools for the processor, including C compilers. Near-term plans are also underway to include a small operating system kernel. These new devices are a hybrid that contains both ASIC and FPLD features. These new large FPLDs with a microprocessor IP core are targeted for system on-a-chip (SOC) applications.

On many of the largest FPLDs, redundant rows of logic elements are included to increase yields. As any VLSI device gets larger the probability of a manufacturing defect increases. If a defective logic element is found during initial testing, the entire row is mapped out and replaced with a spare row of logic elements. This operation is transparent to the user.

3.9 For additional information

This short overview of programmable logic technology has provided a brief introduction to FPLD architectures. Additional CPLD and FPGA manufacturers include Actel, Atmel, Lucent, Quicklogic, and Cypress. Actel, Quicklogic, and Cypress have one-time programmable FPGA devices. These devices utilize antifuse programming technology. Antifuses are open circuits that short circuit or have low impedance only after programming. Trade publications such as *Electronic Design News* periodically have a comparison of the available devices and manufacturers.

Altera MAX 7000, FLEX 10K, and the new APEX and ACEX family data manuals with a more in-depth explanation of device hardware details are available free online at Altera's website, http://www.altera.com.

For further examples of FPGA architectures, details on the Xilinx 4000 or the newer Xilinx Virtex and Virtex II family can be found at http://www.xilinx.com.

An introduction to the mathematics and algorithms used internally by digital logic CAD tools can be found in *Synthesis and Optimization of Digital Circuits* by Giovanni De Micheli, McGraw-Hill, 1994 and *Logic Synthesis and*

Verification Algorithms by Hactel and Somenzi, Kluwer Academic Publishers, 1996.

3.10 Laboratory Exercises

1. Show how the logic equation (A AND NOT(B)) OR (C AND NOT(D)) can be implemented using the following:

   A. The PLA in Figure 3.3
   B. The LUT in Figure 3.9

   Be sure to include the PLA fuse pattern and contents of the LUT.

2. Examine the report file (*.rpt) and use the floorplan editor to explain how the OR-gate design in the tutorial in Chapter 1 was mapped into the FLEX device.

3. Retarget the design from Chapter 1 to the MAX device. Examine the report file (*.rpt) and use the floorplan editor to explain how the OR-gate design in the tutorial in Chapter 1 was mapped into the MAX device.

4. Show how the logic equation (A AND NOT(B)) OR (C AND NOT(D)) can be implemented in the following:

   A. A MAX Logic Element
   B. A FLEX Logic Element
   C. An XC4000 CLB

   Be sure to include the contents of any LUTs required and describe the required mux settings.

5. Using data sheets available on the web, compare and contrast the features of newer generation FPLDs such as Altera's APEX and ACEX and the Xilinx's Virtex and Virtex II families.

CHAPTER 4

Tutorial II: Synchronous Design and Hierarchy

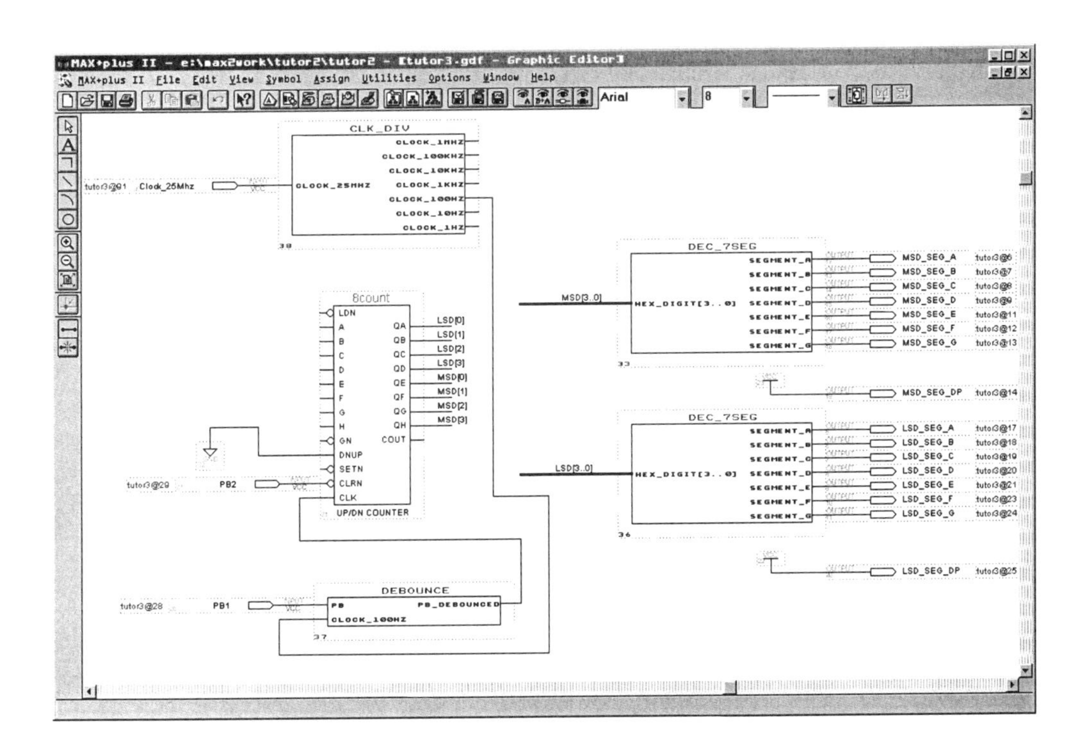

4 Tutorial II: Sequential Design and Hierarchy

The second tutorial contains a more complex design containing sequential logic and hierarchy with a counter and a seven-segment display. To save time, much of the design has already been entered. The existing design will require some modifications. Once completed, you will:

- Understand the fundamentals of hierarchical design tools
- Complete an example of a sequential logic design
- Use the UP1core library designed for the UP 1 and UP 1X boards
- Use seven-segment LED displays, pushbuttons, and the onboard clock
- Use buses in a schematic
- Be able to perform automatic timing analysis of sequential circuits

4.1 Install the Tutorial Files and UP1core Library

Locate the **booksoft\chap4** directory on the CDROM that came with the book. Copy the Chapter 4 tutorial in this directory to your ***drive:*\max2work** directory or a subdirectory. Copy the **booksoft\chap5** directory on the CDROM to your ***drive:*\max2work\UP1core** directory. If the tutorial files or UP1core functions are placed in another directory, it may be necessary to recompile the design since path information for symbols is stored in a project's hierarchical information file (*.hif).

4.2 Open the tutor2 Schematic

Select **File ⇨Open ⇨*drive*:\max2work\tutor2.gdf** and the image in Figure 4.1 should be displayed.

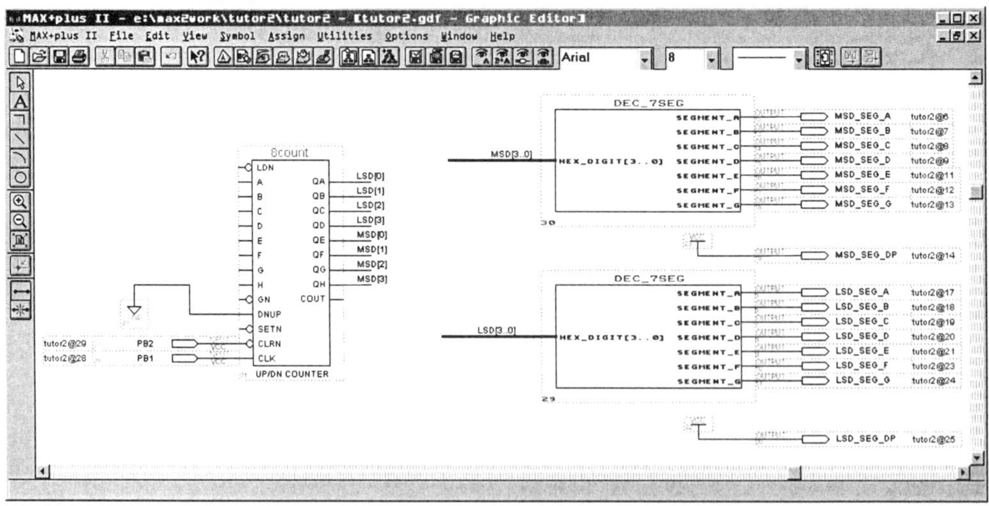

Figure 4.1 The tutor2.gdf schematic.

This design has been partially entered to save time. This is an 8-bit counter design that outputs the counter value to the seven-segment LED displays.

Hit the context sensitive help icon and then select the **8count** symbol to see a help file describing the counter. Examine and close the help file. To output a hexadecimal digit to the seven-segment displays, a seven-segment decoder is required. A seven-segment decoder converts a 4-bit binary value into the seven signals required to drive the seven LED segments in the display. This logic is provided by the dec_7seg UP1core function.

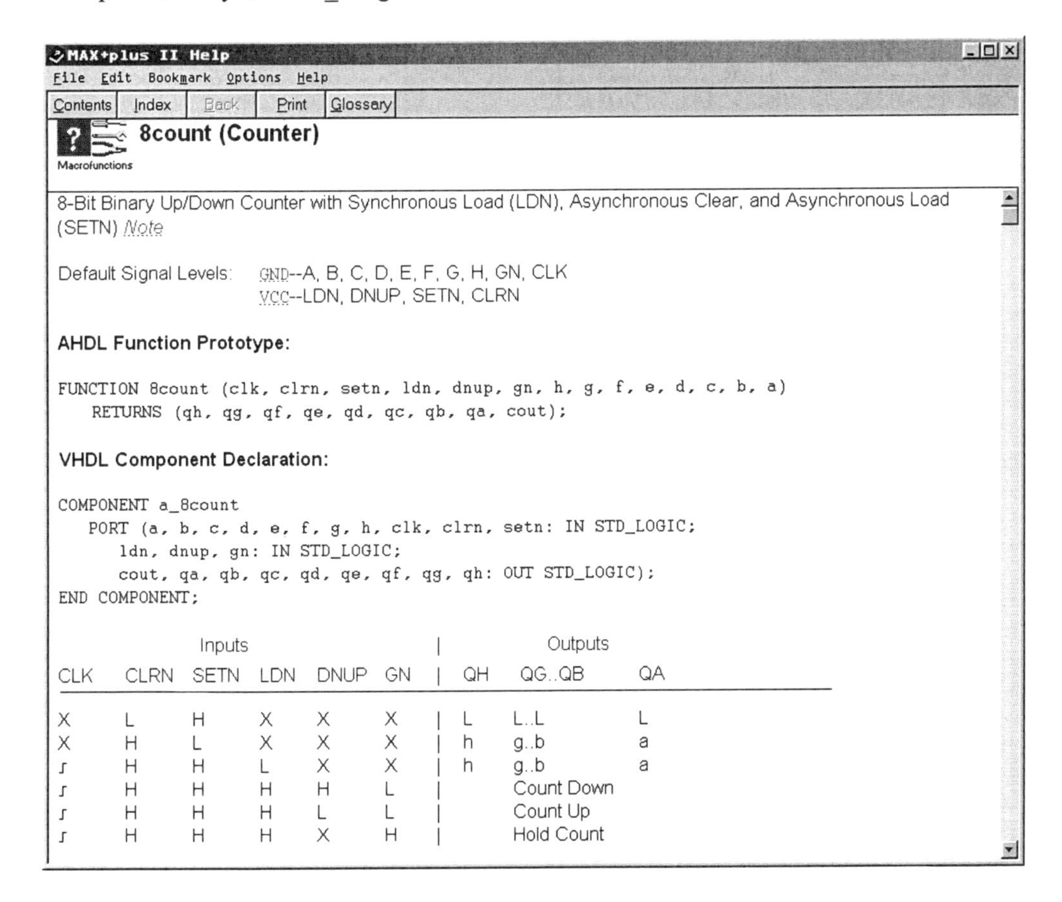

MAX+plus II Help

File Edit Bookmark Options Help

Contents | Index | Back | Print | Glossary

8count (Counter)

Macrofunctions

8-Bit Binary Up/Down Counter with Synchronous Load (LDN), Asynchronous Clear, and Asynchronous Load (SETN) Note

Default Signal Levels: GND--A, B, C, D, E, F, G, H, GN, CLK
VCC--LDN, DNUP, SETN, CLRN

AHDL Function Prototype:

```
FUNCTION 8count (clk, clrn, setn, ldn, dnup, gn, h, g, f, e, d, c, b, a)
   RETURNS (qh, qg, qf, qe, qd, qc, qb, qa, cout);
```

VHDL Component Declaration:

```
COMPONENT a_8count
   PORT (a, b, c, d, e, f, g, h, clk, clrn, setn: IN STD_LOGIC;
      ldn, dnup, gn: IN STD_LOGIC;
      cout, qa, qb, qc, qd, qe, qf, qg, qh: OUT STD_LOGIC);
END COMPONENT;
```

Inputs						Outputs		
CLK	CLRN	SETN	LDN	DNUP	GN	QH	QG..QB	QA
X	L	H	X	X	X	L	L..L	L
X	H	L	X	X	X	h	g..b	a
⨍	H	H	L	X	X	h	g..b	a
⨍	H	H	H	H	L		Count Down	
⨍	H	H	H	L	L		Count Up	
⨍	H	H	H	X	H		Hold Count	

Figure 4.2 8COUNT symbol online help file

Under **Options ⇨ User libraries,** add the path of the UP1core library if it is not already present. The path is ***drive*:\max2work\UP1core**, where *drive* is the drive that contains the MAX+PLUS II tools. The symbols from the UP1core library are now available to be entered in a design. The UP1core functions are special hardware blocks designed to support the easy use of the advanced I/O features found on the UP 1 board. They include pushbuttons, seven-segment displays, keyboard, mouse, and video output.

4.3 Browse the Hierarchy

In engineering, the principle of functional decomposition is normally used in large designs. Complex designs are typically broken into smaller design units. The smaller design units are then more easily understood and implemented. Then the smaller designs are interconnected to form the complex system. The overall design is a hierarchy of interconnected smaller design units. This also promotes the re-use of portions of the design.

The current schematic is a view of the top level of the design. In this design, the problem was decomposed into a design unit or symbol with logic for a counter and another design unit to display the count. Each symbol also has an internal design that can be any combination of another schematic, VHDL, Verilog, or AHDL file.

Double click on the **dec_7seg** symbol to see the underlying VHDL code that describes the internal operation of the dec_7seg block. As shown in Figure 4.3, it is a long case statement that contains the truth table for a hexadecimal to seven-segment decoder. As an alternative, it could be designed at the gate level using this truth table and manually performing logic minimization with seven four-variable Karnaugh maps. Close the VHDL text editor and return to the graphic editor.

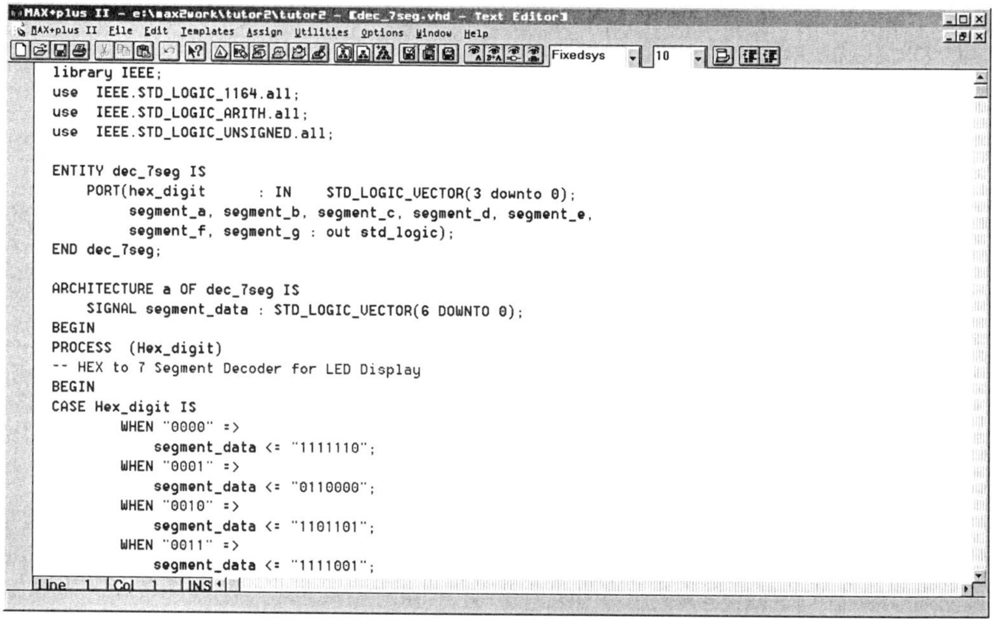

MAX+plus II - e:\max2work\tutor2\tutor2 - [dec_7seg.vhd - Text Editor]

MAX+plus II File Edit Templates Assign Utilities Options Window Help

Fixedsys 10

```
library IEEE;
use  IEEE.STD_LOGIC_1164.all;
use  IEEE.STD_LOGIC_ARITH.all;
use  IEEE.STD_LOGIC_UNSIGNED.all;

ENTITY dec_7seg IS
    PORT(hex_digit      : IN    STD_LOGIC_VECTOR(3 downto 0);
         segment_a, segment_b, segment_c, segment_d, segment_e,
         segment_f, segment_g : out std_logic);
END dec_7seg;

ARCHITECTURE a OF dec_7seg IS
    SIGNAL segment_data : STD_LOGIC_VECTOR(6 DOWNTO 0);
BEGIN
PROCESS  (Hex_digit)
-- HEX to 7 Segment Decoder for LED Display
BEGIN
CASE Hex_digit IS
        WHEN "0000" =>
            segment_data <= "1111110";
        WHEN "0001" =>
            segment_data <= "0110000";
        WHEN "0010" =>
            segment_data <= "1101101";
        WHEN "0011" =>
            segment_data <= "1111001";
```

Line 1 Col 1 INS

Figure 4.3 Internal VHDL code for dec-7seg function.

To see the overall hierarchy of the design, select **MAX+PLUS II ⇨ Hierarchy Display**. As seen in Figure 4.4, the tutor2 schematic is comprised of three symbols. The dec_7seg symbol is used twice in the design and is coded internally in VHDL. The 8Count symbol is defined in terms of another symbol f8count. The f8count design is a *.gdf file or a schematic. In this case, the design hierarchy is three levels deep. After examining the hierarchy display window, close it and return to the graphic editor window that contains the tutor2 schematic. Different versions of the MAX+PLUS II software may contain a slightly different hierarchy display.

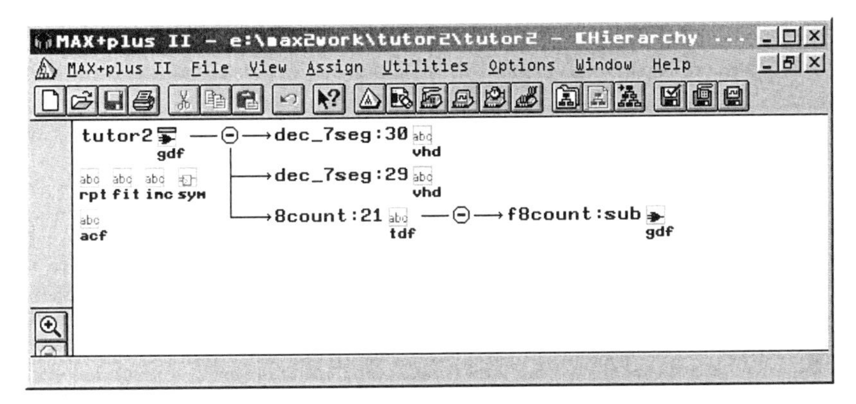

Figure 4.4 Hierarchy display of the tutor2 design.

4.4 Using Buses in a Schematic

Notice the heavy black line feeding into the dec_7seg symbol in Figure 4.5. This is an example of a bus. A bus is just a parallel collection of bits. The bus is labeled MSD[3..0], indicating the bus has four signals, MSD[3], MSD[2], MSD[1], and MSD[0].

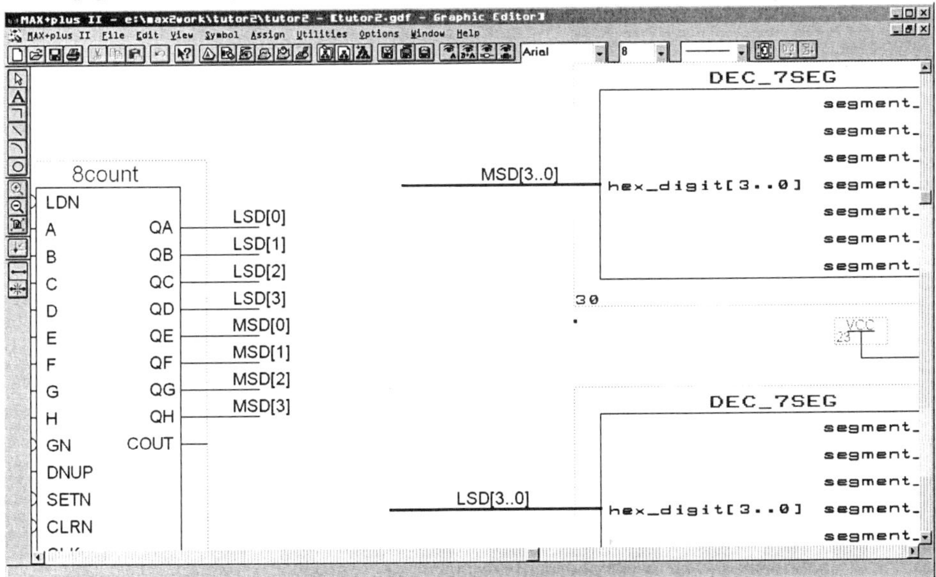

Figure 4.5 Enlarged view of tutor2 design showing bus connnections.

To connect single node lines to a bus it is necessary to assign a name such as MSD[3..0] to the bus. Then the node line that needs to connect to a bus line is given the name of one of the bus elements. As an example, the counter output QH signal line is labeled MSD[3]. To label a signal line or node, right click on the node and select **enter node or bus name**. You can then type in or edit the name. When signal lines have the same name, they are automatically connected in the graphic editor. A physical node line connecting a node and a bus with the same name is optional. Node and bus names must be used when connecting a node to a bus. Examine the counter output node labels and see how they are connected to the dec_7seg seven-segment decoders.

4.5 Testing the Pushbutton Counter and Displays

Compile the design with **File ⇨Project ⇨Save and Compile.**

Select **MAX+PLUS II ⇨ Timing analyzer**. This counter circuit is a sequential design. The primary timing issue in sequential circuits is the maximum clock rate.

Select **Analysis ⇨ Registered Performance** then hit **start**. This runs a timing analysis tool that will determine the maximum clock frequency of the counter circuit. The timing analyzer shows the maximum clock frequency of this logic circuit is approximately 72 Mhz. Clock rates you may obtain will vary depending on the complexity and size of the logic circuits, the speed grade of the chip, and the CAD tool versions and settings. In this design, the clock is supplied by a pushbutton input so a clock input of only a few hertz will be used. Since the UP 1's oscillator is only 25 Mhz this simple counter cannot be overclocked. Close and exit the timing analyzer.

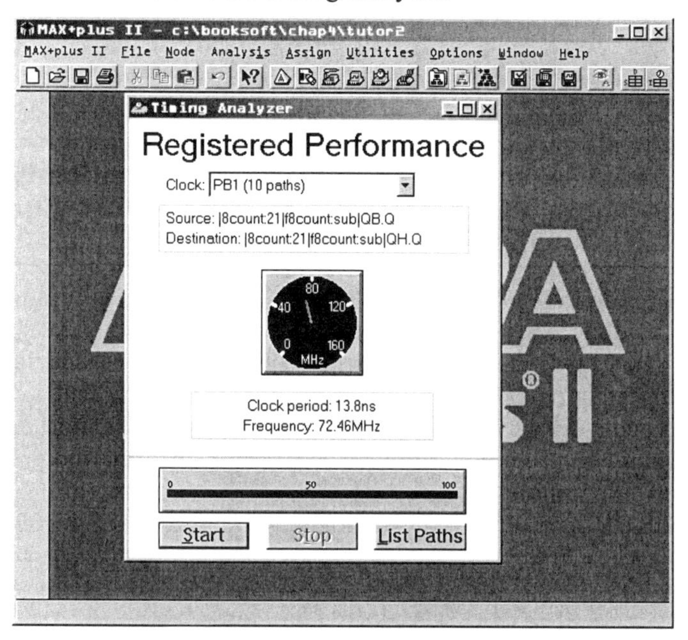

Figure 4.6 Timing analysis using Registered Performance option.

4.6 Testing the Initial Design on the UP 1 Board.

Download the design to the UP 1 board. If you need help downloading the board, refer to section 1.4.

Hit the left FLEX pushbutton several times to clock the counter and watch the seven-segment displays as it counts up. When the pushbutton is hit, it will occasionally count up by more than one. This is a product of mechanical bounce in the pushbutton. The pushbutton contains a metal spring, and it actually makes contact several times before stabilizing. The high-speed logic circuits will react to the switch contact bounce as if several clock signals have occured. This makes the counter count up by more than one.

The actual output of the pushbutton as it appears on a digital oscilloscope is shown in Figure 4.7. When the pushbutton is hit, a random number of pulses appear as the switch contacts bounce several times and then finally stabilize. Several of the pulses will have a voltage and duration long enough to generate extra clock pulses to the counter. An FPLD will respond to pulses in the ns range, and these pulses are in the μs range.

This problem occurs with all pushbuttons in digital designs. If the pushbutton is a DPDT, double-pole double-throw (i.e. has both an ON and an OFF contact), an SR latch is commonly used to remove the contact bounce. The pushbutton on the UP 1 is SPST, single pole single throw, so a time averaging filter is used. This example demonstrates why designs must be tested on actual hardware after simulation. The problem would not have shown up in simulation.

Verify that the right pushbutton resets the counter.

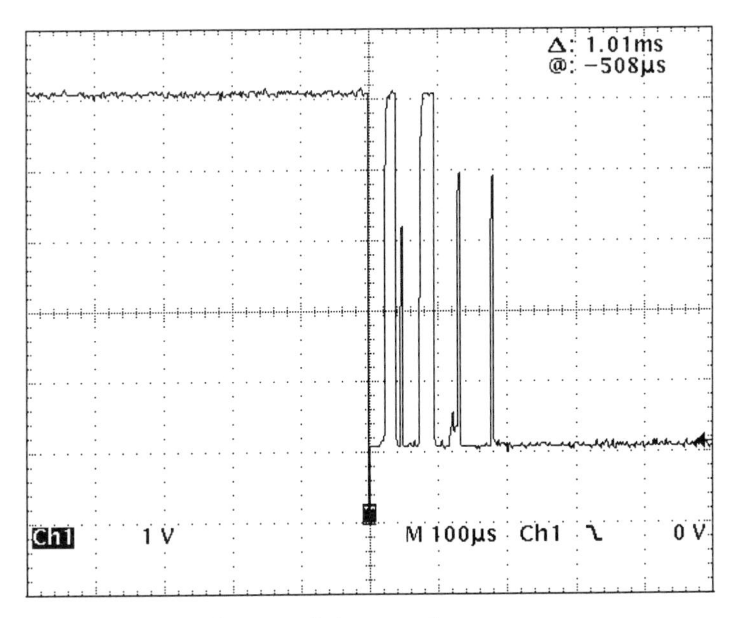

Figure 4.7 Oscillosope display of pushbutton switch contact bounce.

4.7 Fixing the Switch Contact Bounce Problem

For the hardware implementation to work correctly, the switch contact bounce must be removed. A logic circuit that filters the pushbutton output using a small shift register can be used to debounce the output. This process is called switch debouncing. Add the symbol **debounce** from the UP1core library to the schematic.

The internal VHDL design in the debounce module generates the switch debounce circuit. The debounce circuit contains a 4-bit shift register that is clocked at 100Hz. The shift register shifts in the inverted pushbutton output. When any of the 4-bits of the shift register (i.e. four 10 ms. time spaced samples of the pushbutton's output) are high the output of the debounce circuit changes to high. When all 4-bits of the shift register are low the output goes low. This delays the high to low change until after the contact bounce stops.

Disconnect the pushbutton from the 8COUNT CLK pin, connect it to the pushbutton input pin to PB on the debounce symbol. Now connect the PB_DEBOUNCED pin to the 8COUNT CLK pin. The debounce circuit needs a 100Hz clock signal for the time averaging filter. The clock needed is much slower than the 25Mhz system clock, so a clock prescalar is needed. A clock prescalar is a logic circuit that divides a clock signal.

Add the **clk_div** symbol from the UP 1 core library to the schematic. Connect the 100Hz input pin on the debounce symbol to the 100Hz output pin on the clock prescalar. Connect the 25Mhz clock input on the **clk_div** symbol to an input pin. Label the pin **clock_25Mhz** and assign pin 91 to the input pin. Pin 91 is connected to a 25Mhz oscillator on the UP 1 board.

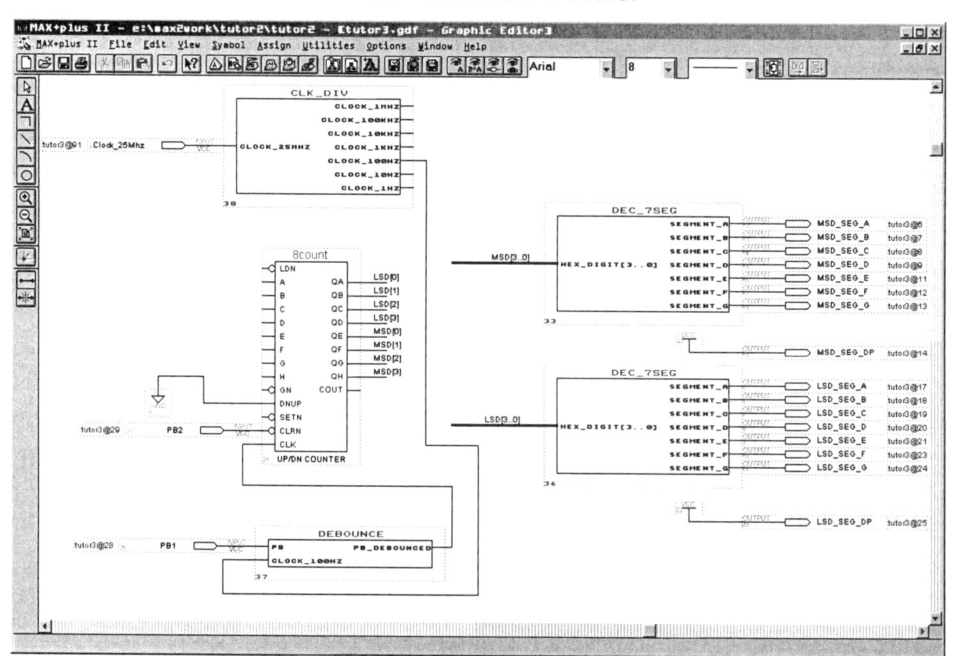

Figure 4.8 Modified tutor2 design schematic.

4.8 Testing the Modified Design on the UP 1 Board.

Verify that your schematic has the same connections as seen in Figure 4.8. Compile the design and download the design to the UP 1 board. Hit the left FLEX pushbutton several times to clock the counter and watch the seven-segment displays as it counts up. It should now count only once when the pushbutton is hit. The UP1core functions dec_7seg, clk_div, and debounce will be useful in future design projects using the UP 1 board. They can be used in any VHDL, Verilog, schematic, or AHDL designs by using the graphical editor and UP1core symbols.

4.9 Laboratory Exercises

1. Simulate the initial design without the switch debounce circuit by setting up an initial reset pulse and a periodic 200 ns clock input in the simulator. In sequential simulations, turn on the **check setup and hold** option before running the simulator. This will check for flip-flop timing problems that would otherwise go undetected in the simulation.

2. Modify the counter circuit so that it counts down or up depending on the state of FLEX DIP switch 1. FLEX DIP switch 1 is connected to pin 41.

3. Modify the counter circuit so that it parallel loads a count value from the eight FLEX DIP switches on the UP 1 board when PB2 is pushed. Since the FLEX DIP switch inputs are only used when PB2 is hit, they do not need to be debounced. Here are the pin connections for the FLEX DIP switches.

Input Pin	Pin
FLEX_switch_1	41
FLEX_switch_2	40
FLEX_switch_3	39
FLEX_switch_4	38
FLEX_switch_5	36
FLEX_switch_6	35
FLEX_switch_7	34
FLEX_switch_8	33
(1 = Open, 0 = Closed)	

4. Build a stopwatch with the following modifications to the design. Disconnect the counter clk line and connect it to the clock_10hz pin on the clock_div symbol. Clock a toggle flip-flop with the pb_debounced output. A toggle flip-flop, tff, can be found in the prim symbol library. A toggle flip-flop's output changes state every time it is clocked. Connect the output of the toggle flip-flop to the GN line on the counter. The count should start and stop when PB1 is hit. Elapsed time in tenths of seconds should be displayed in hexadecimal. Pushing PB2 should reset the stopwatch.

5. Modify the counter or stopwatch circuit so that it uses the MAX chip instead of the FLEX chip. Compare the maximum clock rate on the FLEX chip to that of the MAX chip. A table of MAX chip pin assignments can be found in Appendix C.

6. The elapsed time in the stopwatch from problem 3 is displayed in hexadecimal. Replace the counter with two cascaded binary-coded-decimal (BCD) counters so that it displays the elapsed time as two decimal digits.

7. Build a watch by expanding the counter circuit to count seconds, hours, and minutes. The two pushbuttons reset and start the watch. Use input from the FLEX DIP switch to select the different outputs (hours, minutes or seconds) that will appear on the seven-segment displays.

8. Retarget the example design to the MAX chip. Pin numbers for the MAX decimal point LED can be found in Appendix C. It will be necessary to connect jumper wires from the MAX header to the pushbuttons. Select pins near the pushbuttons to simplify wiring the jumpers.

9. Replace the 8count logic with a VHDL counter design, simulate the design, and verify operation on the UP 1 board. Read Chapter 5 and note the example counter design in section 6.10.

10. Draw a schematic, devclop a simulation, and download a design to the UP 1 board that uses the LED displays for outputs and the DIP switch for input, to test the following functions found in the MF library:

8FADD:	wire B input to 00001010.
8MCOMP:	wire B input to 00010000.
8DFF:	latch FLEX DIP switch input
BARRELST:	wire S0-S2 and A-E to FLEX DIP switch input.

Use the FLEX DIP switch to provide eight inputs. Connect each input to a D flip-flop and display the output in hex on the two seven-segment displays. Use a debounced pushbutton input for the clock. Use the second pushbutton for a Clear or Reset input.

11. Draw a schematic, develop a simulation, and download a design to the UP 1 board to test the following functions found in the MEGA_LPM library:

LPM_ADD_SUB:	a 4-bit adder/subtractor; test the add operation
LPM_ADD_SUB:	a 4-bit adder/subtractor; test the subtract operation
LPM_COMPARE:	compare two 4-bit unsigned numbers
LPM_DECODE:	a 4 to 16-bit decoder
LPM_CLSHIFT:	a 4-bit shift register
LPM_MULT:	a 4-bit unsigned multiply

The LPM megafunctions require several parameters to specify bus size and other various options. For this problem, do not use pipelining and use the unregistered input options. Refer to the online help files for each LPM function for additional information. In the **enter symbol** window, use the megawizard button to help configure LPM symbols. Use the FLEX DIP switch for eight inputs as needed and display the output in hex on the two seven-segment displays. Use a debounced pushbutton input for the clock, if one is required. Use the second pushbutton for a Clear or Reset input. Use the timing analyzer to determine the worst-case delay time for each function.

12. Draw a schematic and develop a simulation to test the LPM_ROM megafunction. Create a sixteen word ROM with eight data bits per word. Specify initial values in hex for the ROM in a memory initialization file (*.mif) file. The contents of each memory location should be initialized to four times its address. See MIF in the online help for details on the syntax of a MIF file. Enter the address in four FLEX DIP switches and display the data from the ROM in the two seven-segment LEDs. Determine the access time of the ROM.

13. Using gates and the DFF part from the prim library, design a circuit that implements the state machine shown below. Use two D flip-flops with an encoded state. For the encoded states use A = "00", B = "01", and C = "10". Ensure that the undefined "11" state enters a known state. Enter the design using the graphical editor. Develop a simulation that tests the state machine for correct operation. The simulation should test all states and arcs in the state diagram and the "11" state. Use the timing analyzer's **Analysis ⇨Registered Performance** option to determine the maximum clock frequency on the FLEX device. Use an asynchronous reset.

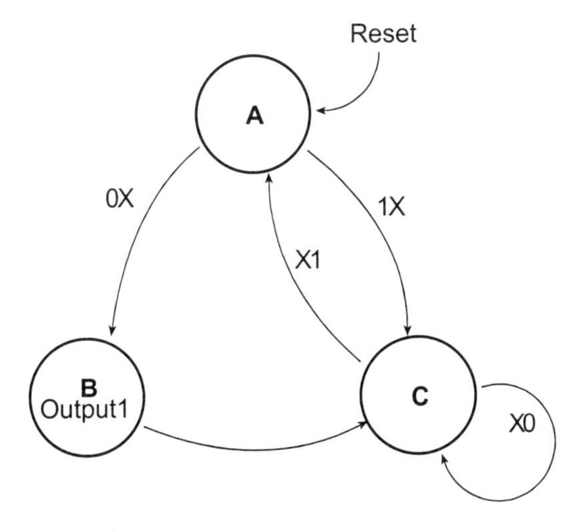

14. Repeat the previous problem using one-hot encoding. Recall that one-hot encoding uses one flip-flop per state, and only one flip-flop is ever active at any given time. The state encoding for the one-hot state machine would be A = "100", B = "010", and C = "001".

Start with a reset in the simulation. It is not necessary to test illegal states in the one-hot simulation. One-hot state machine encoding is recommended by many FPLD device manufacturers.

CHAPTER 5

UP1core Library Functions

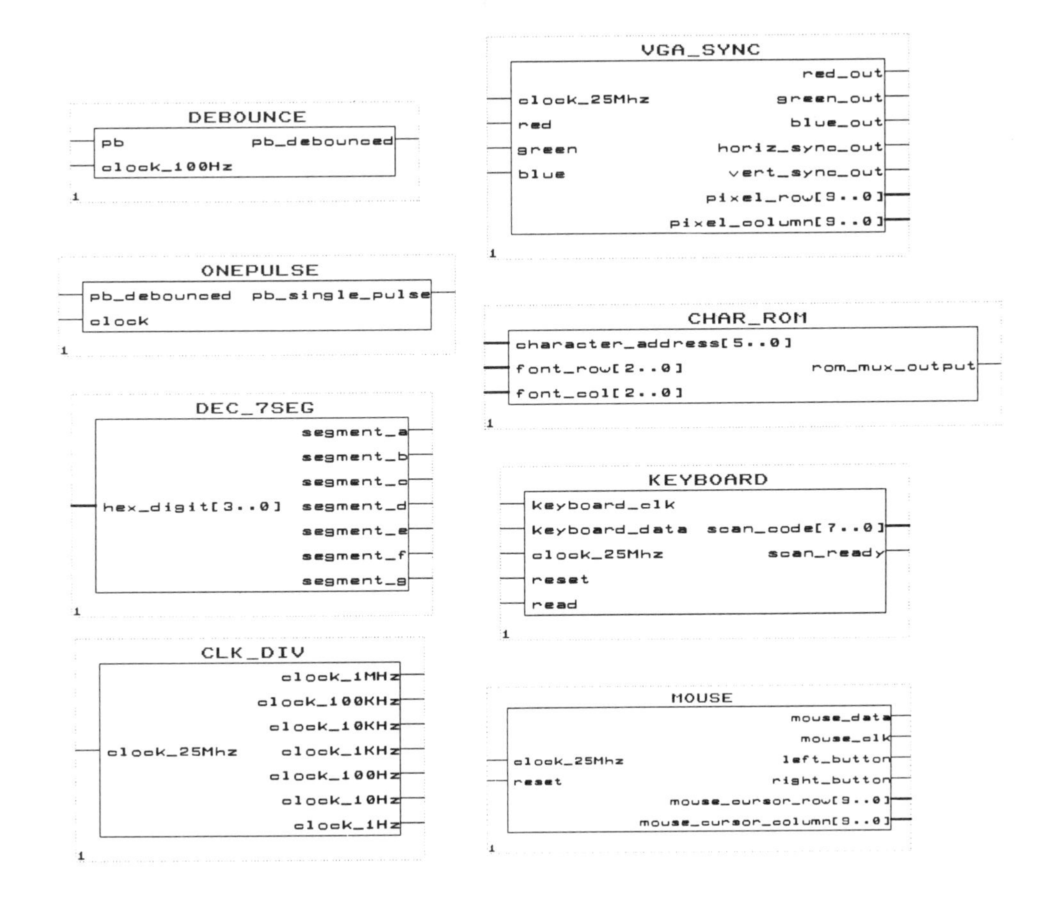

5 UP1core Library Functions

In complex hierarchical designs, intellectual property (IP) cores are frequently used. An IP core is a previously developed synthesizable hardware design that provides a widely used function. Commercially licensed IP cores include functions such as microprocessors, microcontrollers, bus interfaces, multimedia and DSP functions, and communications controllers. IP cores promote design reuse and reduce development time by providing common hardware functions for use in a new design.

The UP1core functions listed in Table 5.1 are designed to simplify use of the UP 1 board's switch, keyboard, mouse, LED display, and video output features. They can be used in schematic capture, VHDL, AHDL, or Verilog based designs.

Table 5.1 The UP1core Functions.

UP1core Name	Description
DEC_7SEG	Hexadecimal to Seven-segment Decoder for the UP 1's LED Displays
DEBOUNCE	Pushbutton Debounce Circuit
ONEPLUSE	Pushbutton Single Pulse Circuit
CLK_DIV	25Mhz Clock Prescaler with 7 frequency outputs (1Mhz to 1hz)
VGA_SYNC	VGA Sync signal generator for UP 1 that outputs pixel addresses
CHAR_ROM	Character Font ROM for video character generation
KEYBOARD	Reads keyboard scan codes from the UP 1's PS/2 connector
MOUSE	Reads mouse data and outputs cursor row and column address

UP1cores can be used as symbols from the UP1core library, accessed via a VHDL package, or used as a component in other VHDL files. An example of using the UP1core package in VHDL can be found in the file, UP1pack.vhd. The use of UP1pack's VHDL package saves repeating lengthy component declarations in VHDL-based designs.

This section contains a one-page summary of each UP1core interface. VHDL source code is provided for all UP1cores on the CDROM. Additional documentation, examples, and interface details can be found in later chapters on video signal generation, the keyboard, and the mouse. The clk_div, dec_7seg, and debounce functions were used in the tutorial design example in Chapter 4.

For correct operation of the UP1core functions, I/O pin assignments must be made as shown in the description of each UP1core function. Clock inputs are also required on several of the UP1core functions. The Clk_Div UP1core is setup to provide the clock signals needed. Source code for the UP1core functions must be in the user library search path or in the project directory. Review section 4.2 for additional information on checking the library path.

5.1 UP1core DEC_7SEG: Hex to Seven-segment Decoder

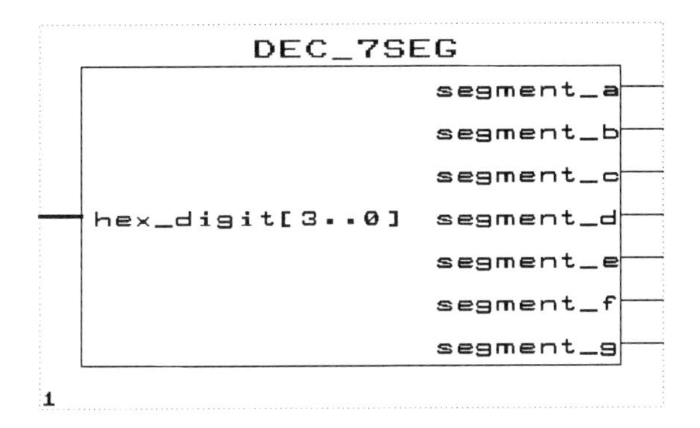

Figure 5.1 Symbol for DEC_7SEG UP1core.

The UP1core Dec_7seg shown in Figure 5.1 is a hexadecimal to seven-segment display decoder with active low outputs. This function is used to display hex numbers in the UP 1's seven-segment LED displays.

5.1.1 VHDL Component Declaration

```
COMPONENT dec_7seg
	PORT( hex_digit		: IN	STD_LOGIC_VECTOR ( 3 DOWNTO 0 );
		seg_a, seg_b, seg_c,
		seg_d, seg_e, seg_f,
		seg_g		: OUT	STD_LOGIC );
END COMPONENT;
```

5.1.2 Inputs

Hex_digit is the 4-bit hexadecimal value to send to the LED display. Hex_digit[3] is the most-significant bit.

5.1.3 Outputs

Segments a through g are active low and should be connected as output pins to the corresponding pin on a seven-segment display. See Chapter 2 for seven-segment display pin connections on the UP 1 board.

5.2 UP1core Debounce: Pushbutton Debounce

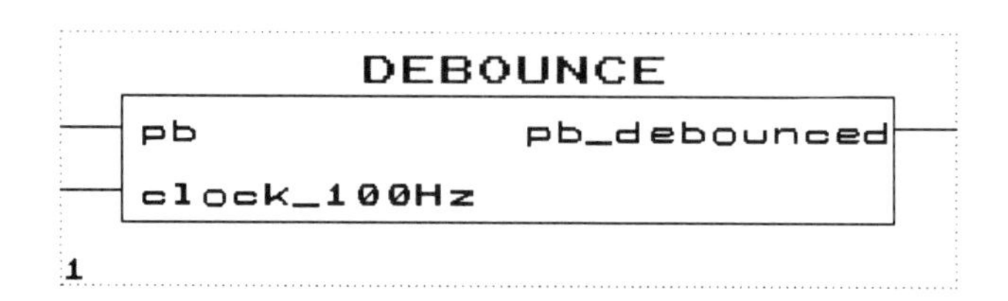

Figure 5.2 Symbol for DEBOUNCE UP1core.

The UP1core Debounce shown in Figure 5.2 is a pushbutton debounce circuit. This function is used to filter mechanical bounce on the UP 1's pushbuttons. A shift register is used to filter switch contact bounce. The shift register takes several time samples of the pushbutton input and changes the output only after several sequential samples are the same value.

5.2.1 VHDL Component Declaration

```
COMPONENT debounce
        PORT( pb, clock_100Hz        : IN        STD_LOGIC;
              pb_debounced           : OUT       STD_LOGIC;
END COMPONENT;
```

5.2.2 Inputs

PB is the raw pushbutton input. It should be tied to an input pin connected to a pushbutton. See Chapter 2 for MAX and FLEX pushbutton pin numbers.

Clock is a clock signal of approximately 100Hz that is used for the internal 50ms switch debounce filter circuits.

5.2.3 Outputs

PB_debounced is the debounced pushbutton output. The output will remain low until the pushbutton is released.

5.3 UP1core OnePulse: Pushbutton Single Pulse

ONEPULSE
pb_debounced pb_single_pulse
clock
1

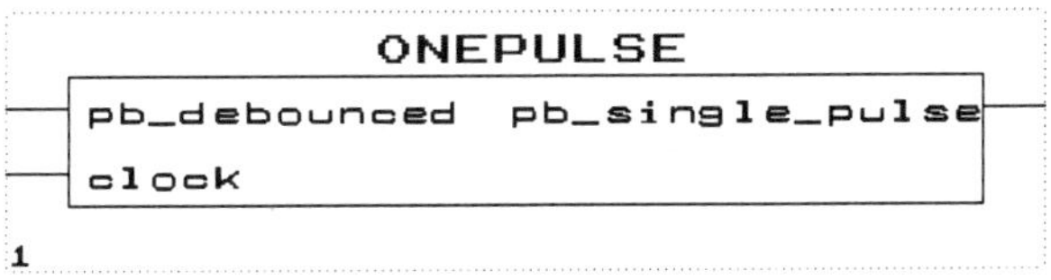

Figure 5.3 Symbol for ONEPULSE UP1core.

The UP1core Onepulse shown in Figure 5.3 is a pushbutton single pulse circuit. Output from the pushbutton is High for only one clock cycle no matter how long the pushbutton is depressed. This function is useful in state machines that read external pushbutton inputs. In general, fewer states are required when inputs only activate for one clock cycle. Internally, an edge-triggered flip-flop is used to build a simple state machine.

5.3.1 VHDL Component Declaration

```
COMPONENT onepulse
        PORT( pb_debounced, clock   : IN    STD_LOGIC;
              pb_single_pulse       : OUT   STD_LOGIC );
END COMPONENT;
```

5.3.2 Inputs

Pb_debounced is the debounced pushbutton input. It should be connected to a debounced pushbutton.

Clock is the user's state-machine clock. It can be any frequency.

5.3.3 Outputs

PB_single_pulse is the output, which is High for only one clock cycle when a pushbutton is hit.

5.4 UP1core Clk_Div: Clock Divider

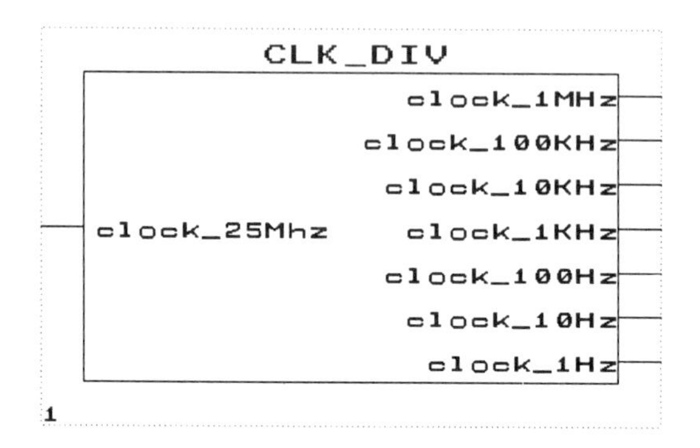

Figure 5.4 Symbol for CLK_DIV UP1core.

The UP1core CLK_DIV shown in Figure 5.4 is used to provide clock signals slower than the on-board 25.175Mhz oscillator. These signals are obtained by dividing down the 25.175 clock input signal. Multiple output taps provide clock frequencies in powers of ten.

5.4.1 VHDL Component Declaration

```
COMPONENT clk_div
        PORT(   clock_25Mhz     : IN    STD_LOGIC;
                clock_1MHz      : OUT   STD_LOGIC;
                clock_100KHz    : OUT   STD_LOGIC;
                clock_10KHz     : OUT   STD_LOGIC;
                clock_1KHz      : OUT   STD_LOGIC;
                clock_100Hz     : OUT   STD_LOGIC;
                clock_10Hz      : OUT   STD_LOGIC;
                clock_1Hz       : OUT   STD_LOGIC );
END COMPONENT;
```

5.4.2 Inputs

Clock_25Mhz is an input pin that should be connected to the on-board 25.175Mhz oscillator. Pin numbers for the 25.175Mhz clock are 91 for the FLEX chip and 83 for the MAX chip.

5.4.3 Outputs

Clock_1Mhz through clock_1Hz provide output signals of the appropriate frequency. The actual frequency is 1.007 $\pm$ .005% times the listed value.

5.5 UP1core VGA_Sync: VGA Video Sync Generation

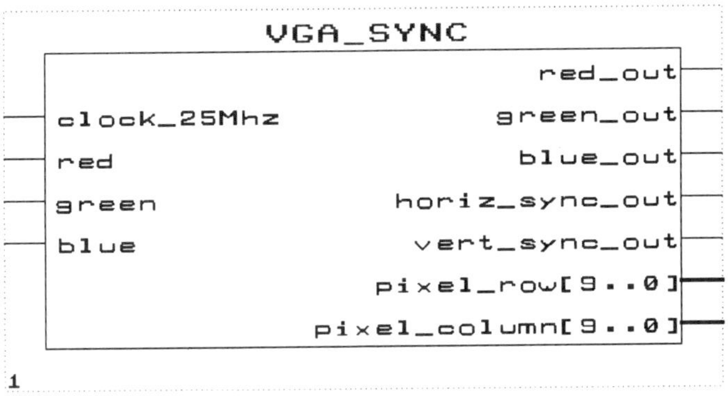

VGA_SYNC
clock_25Mhz
red
green
blue
red_out
green_out
blue_out
horiz_sync_out
vert_sync_out
pixel_row[9..0]
pixel_column[9..0]
1

Figure 5.5 Symbol for VGA_SYNC UP1core.

The UP1core VGA_Sync shown in Figure 5.5 provides horizontal and vertical sync signals to generate an eight-color 640 by 480 pixel VGA video image. Video can only be generated on the FLEX chip. For more detailed information on video signal generation see Chapter 9.

5.5.1 VHDL Component Declaration

```
COMPONENT vga_sync
        PORT( clock_25Mhz, red, green, blue      : IN STD_LOGIC;
        red_out, green_out, blue_out,
        horiz_sync_out, vert_sync_out            : OUT  STD_LOGIC;
        pixel_row, pixel_column         :OUT STD_LOGIC_VECTOR( 9 DOWNTO 0 ) );
END COMPONENT;
```

5.5.2 Inputs

Clock_25Mhz is an input pin that must be connected to the on-board 25.175Mhz oscillator. The pin number for the 25.175Mhz clock input is 91 on the FLEX chip.

Red, Green, and Blue inputs provide the color information for the video signal. External user logic must generate the RGB signals.

5.5.3 Outputs

Horiz_sync is an output pin that should be tied to the horizontal sync, pin 240, on the FLEX chip.

Vert_sync is an output pin that should be tied to the vertical sync, pin 239, on the FLEX chip.

Red_out is an output pin that should be tied to the red signal, pin 236, on the FLEX chip.

Green_out is an output pin that should be tied the green signal, pin 237, on the FLEX chip.

Blue_out is an output pin that should be tied to the blue signal, pin 238, on the FLEX chip.

Pixel_row and Pixel_column are outputs that provide the current pixel address. These outputs are used by user logic to generate RGB color input data.

5.6 UP1core CHAR_ROM: Character Generation ROM

CHAR_ROM
character_address[5..0]
font_row[2..0]
font_col[2..0]
rom_mux_output
1

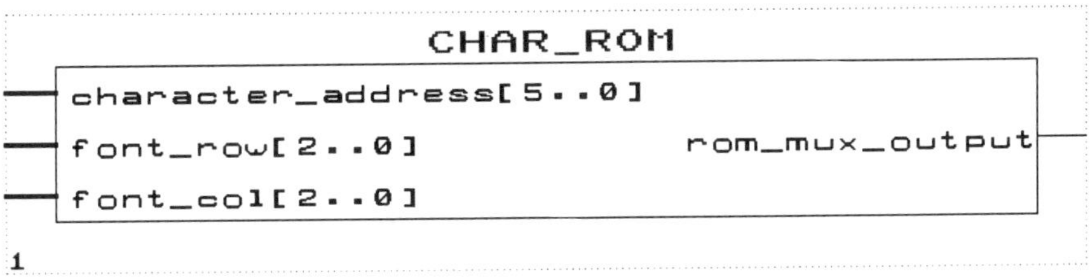

Figure 5.6 Symbol for CHAR_ROM UP1core.

The UP1core Char_ROM shown in Figure 5.6 is a character generation ROM used to generate text in a video display. Each character is represented by an 8 by 8 pixel font. For more information on video character generation see Chapter 9. Character codes are listed in Table 9.1 of section 9.9. Font data is contained in the memory initialization file, tcgrom.mif. Two FLEX EABs are required for the ROM that holds the font data.

5.6.1 VHDL Component Declaration

```
COMPONENT char_rom
        PORT( character_address     : IN STD_LOGIC_VECTOR( 5 DOWNTO 0 );
              font_row, font_col    : IN STD_LOGIC_VECTOR( 2 DOWNTO 0 );
              rom_mux_output        : OUT STD_LOGIC);
END COMPONENT;
```

5.6.2 Inputs

Character_address is the address of the alphanumeric character to display.

Font_row and font_col are used to index through the 8 by 8 font to address a single pixel for video generation.

5.6.3 Outputs

Rom_mux_output is the pixel font value indexed by the address inputs. It is used by user logic to generate the RGB pixel color data for the video signal.

5.7 UP1core Keyboard: Read Keyboard Scan Code

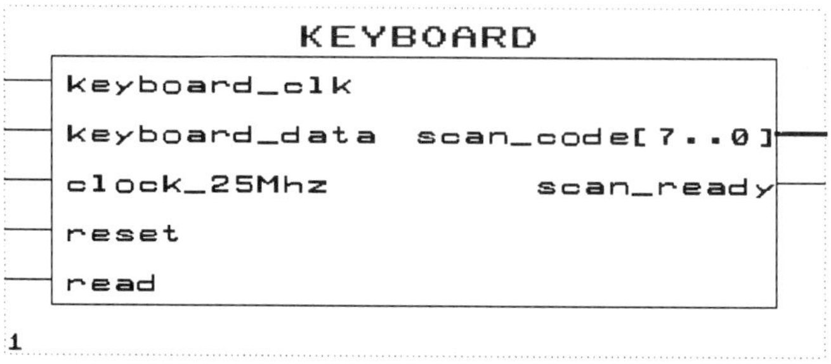

KEYBOARD
keyboard_clk
keyboard_data scan_code[7..0]
clock_25Mhz scan_ready
reset
read
1

Figure 5.7 Symbol for KEYBOARD UP1core.

The UP1core Keyboard shown in Figure 5.7 is used to read the PS/2 keyboard scan code from a keyboard attached to the FLEX chip PS/2 connector. This function converts the serial data from the keyboard to parallel format to produce the scan code output. For detailed information on keyboard applications and scan codes see Table 10.2 in Chapter 10.

5.7.1 VHDL Component Declaration

```
COMPONENT keyboard
        PORT( keyboard_clk, keyboard_data, clock_25Mhz ,
              reset, read      : IN STD_LOGIC;
              scan_code        : OUT STD_LOGIC_VECTOR( 7 DOWNTO 0 );
              scan_ready       : OUT STD_LOGIC);
END COMPONENT;
```

5.7.2 Inputs

Clock_25Mhz is an input pin that must be connected to the on-board 25.175Mhz oscillator. The pin number for the 25.175Mhz clock is 91 on the FLEX chip.

Keyboard_clk and keyboard_data are input data lines from the keyboard. Keyboard_clk is pin 30 and keyboard_data is pin31.

Read is a handshake-input signal. The rising edge of the read signal clears the scan ready signal. Reset is an input that clears the internal registers and flags used for serial-to-parallel conversion.

5.7.3 Outputs

Scan_code contains the bytes transmitted by the keyboard when a key is pressed or released. See Table 10.2 in Chapter 10 for a listing of scan codes. Scan codes for a single key are a sequence of several bytes. A make code is sent when a key is hit, and a break code is sent whenever a key is released.

Scan_ready is a handshake output signal that goes High when a new scan code is sent by the keyboard. The read input clears scan_ready. The scan_ready handshake line should be used to ensure that a new scan code is read only once.

5.8 UP1core Mouse: Mouse Cursor

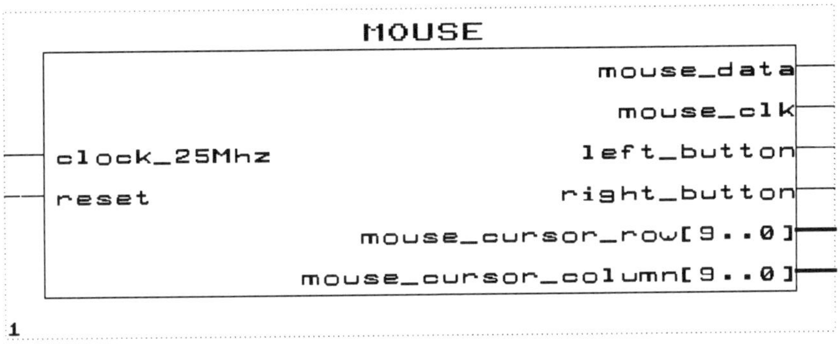

```
MOUSE
clock_25Mhz                      mouse_data
reset                            mouse_clk
                                 left_button
                                 right_button
                                 mouse_cursor_row[9..0]
                                 mouse_cursor_column[9..0]
1
```

Figure 5.8 Symbol for MOUSE UP1core.

The UP1core Mouse shown in Figure 5.8 is used to read position data from a mouse attached to the FLEX chip PS/2 connector. It outputs a row and column cursor address for use in video applications. The mouse must be attached to the UP 1 board prior to downloading for proper initialization. Detailed information on mouse applications, commands, and data formats can be found in Chapter 11.

5.8.1 VHDL Component Declaration

```
COMPONENT mouse
        PORT( clock_25Mhz, reset        : IN       STD_LOGIC;
              mouse_data                : INOUT    STD_LOGIC;
              mouse_clk                 : INOUT    STD_LOGIC;
              left_button, right_button : OUT      STD_LOGIC;
              mouse_cursor_row          : OUT STD_LOGIC_VECTOR( 9 DOWNTO 0 ) );
              mouse_cursor_column       : OUT STD_LOGIC_VECTOR( 9 DOWNTO 0 ) );
END COMPONENT;
```

5.8.2 Inputs

Clock_25Mhz is an input pin that must be connected to the on-board 25.175Mhz oscillator. The pin number for the 25.175Mhz clock is 91 on the FLEX chip.

Mouse_clk and mouse_data are bi-directional data lines from the mouse. Mouse_clk is pin 30 and mouse_data is pin 31.

5.8.3 Outputs

Mouse_cursor_row and mouse_cursor_column are outputs which contain the current address of the mouse cursor in the 640 by 480 screen area. The cursor is initialized to the center of the screen. Left_button and right_button outputs are High when the corresponding mouse button is pressed.

CHAPTER 6

Using VHDL for Synthesis of Digital Hardware

```
ARCHITECTURE behavior OF ALU IS
SIGNAL temp_output: STD_LOGIC_VECTOR( 7 DOWNTO 0 );
BEGIN
PROCESS (Op_code, A_input, B_input)
        BEGIN
        -- Select Arithmetic/Logical Operation
        CASE Op_Code ( 2 DOWNTO 1 ) IS
                WHEN "00" =>
                   temp_output <= A_input + B_input;
                WHEN "01" =>
                   temp_output <= A_input - B_input;
                WHEN "10" =>
                   temp_output <= A_input AND B_input;
                WHEN "11" =>
                   temp_output <= A_input OR B_input;
                WHEN OTHERS =>
                   temp_output <= "00000000";
        END CASE;
        -- Select Shift Operation
        IF Op_Code( 0 ) = '1' THEN
                Alu_output <= temp_output( 6 DOWNTO 0 ) & '0';
        ELSE
                Alu_output <= temp_output;
        END IF;
END PROCESS;
END behavior;
```

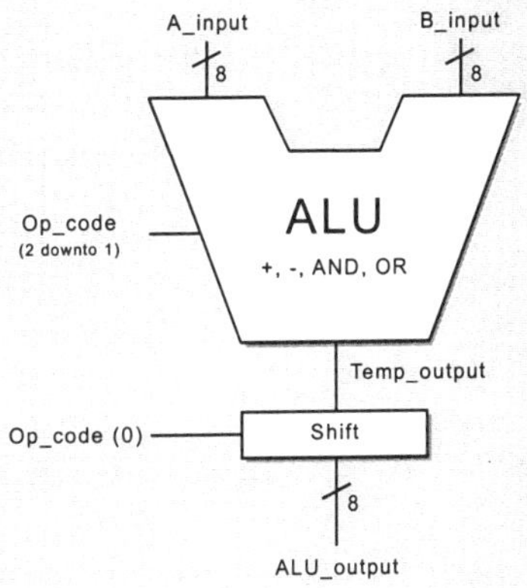

6 Using VHDL for Synthesis of Digital Hardware

In the past, most digital designs were manually entered into a schematic entry tool. With increasingly large and more complex designs, this is a tedious and time-consuming process. Logic synthesis using hardware description languages is becoming widely used since it greatly reduces development time and cost. It also enables more exploration of design alternatives, more flexibility to changes in the hardware technology, and promotes design reuse.

VHDL is a language widely used to model and design digital hardware. VHDL is the subject of IEEE standards 1076 and 1164 and is supported by numerous CAD tool and programmable logic vendors. VHDL is an acronym for VHSIC Hardware Description Language. VHSIC, Very High Speed Integrated Circuits, was a USA Department of Defense program in the 1980s that sponsored the early development of VHDL. VHDL has syntax similar to ADA and PASCAL.

Conventional programming languages are based on a sequential operation model. Digital hardware devices by their very nature operate in parallel. This means that conventional programming languages cannot accurately describe or model the operation of digital hardware since they are based on the sequential execution of statements. VHDL is designed to model parallel operations.

IT IS CRUCIAL TO REMEMBER THAT VHDL MODULES, CONCURRENT STATEMENTS, AND PROCESSES ALL OPERATE IN PARALLEL.

In VHDL, variables change without delay and signals change with a small delay. For VHDL synthesis, signals are normally used instead of variables so that simulation works the same as the synthesized hardware.

A subset of VHDL is used for logic synthesis. In this section, a brief introduction to VHDL for logic synthesis will be presented. It is assumed that the reader is already familiar with basic digital logic devices and PASCAL, ADA, or VHDL.

Whenever you need help with VHDL syntax, VHDL templates of common statements are available in the MAX+PLUS II online help. In the text editor, just click the right mouse button and select VHDL Templates.

6.1 VHDL Data Types

In addition to the normal language data types such as Boolean, integer, and real, VHDL contains new types useful in modeling digital hardware. For logic synthesis, the most important type is standard logic. Type standard logic, STD_LOGIC, is normally used to model a logic bit. To accurately model the operation of digital circuits, more values than "0" or "1" are needed for a logic bit. In the logic simulator, a standard logic bit can have nine values, U, X, 0, 1, Z, W, L, H, and "-". U is uninitialized and X is forced unknown. Z is tri-state or high impedance. L and H are weak "0" and weak "1". "-" is don't care. Type STD_LOGIC_VECTOR contains a one-dimensional array of

STD_LOGIC bits. Using these types normally requires the inclusion of special standard logic libraries at the beginning of each VHDL module. The value of a standard logic bit can be set to '0' or '1' using single quotes. A standard logic vector constant, such as the 2-bit zero value, "00" must be enclosed in double quotes.

6.2 VHDL Operators

Table 6.1 lists the VHDL operators and their common function in VHDL synthesis tools.

Table 6.1 VHDL Operators.

VHDL Operator	Operation
+	Addition
-	Subtraction
*	Multiplication*
/	Division*
MOD	Modulus*
REM	Remainder*
&	Concatenation – used to combine bits
SLL**	logical shift left
SRL**	logical shift right
SLA**	arithmetic shift left
SRA**	arithmetic shift right
ROL**	rotate left
ROR**	rotate right
=	equality
/=	Inequality
<	less than
<=	less than or equal
>	greater than
>=	greater than or equal
NOT	logical NOT
AND	logical AND
OR	logical OR
NAND	logical NAND
NOR	logical NOR
XOR	logical XOR
XNOR*	logical XNOR

*Not supported in many VHDL synthesis tools. In the MAX+PLUS II tools, only multiply and divide by powers of two (shifts) are supported. Mod and Rem are not supported in MAX+PLUS II. Efficient design of multiply or divide hardware typically requires the user to specify the arithmetic algorithm and design in VHDL.

** Supported only in 1076-1993 VHDL.

Table 6.2 illustrates several useful conversion functions for type STD_LOGIC and integer:

Table 6.2 STD_LOGIC conversion functions.

Function	Example:
TO_STDLOGICVECTOR(*bit_vector*)	**TO_STDLOGICVECTOR(** X"FFFF")
Converts a bit vector to a standard logic vector.	Generates a 16-bit standard logic vector of ones. "X" indicates hexadecimal and "B" is binary.
CONV_STD_LOGIC_VECTOR(*integer, bits*)	**CONV_STD_LOGIC_VECTOR(** 7, 4)
Converts an integer to a standard logic vector.	Produces a standard logic vector of "0111".
CONV_INTEGER(*std_logic_vector*)	**CONV_INTEGER(** "0111")
Converts a standard logic vector to an integer.	Produces an integer value of 7.

6.3 VHDL Based Synthesis of Digital Hardware

VHDL can be used to construct models at a variety of levels such as structural, behavioral, register transfer level (RTL), and timing. An RTL model of a circuit described in VHDL describes the input/output relationship in terms of dataflow operations on signal and register values. If registers are required, a synchronous clocking scheme is normally used. Sometimes an RTL model is also referred to as a dataflow-style model.

VHDL simulation models often include physical device time delays. In VHDL models written for logic synthesis, timing information should not be provided.

For timing simulations, the CAD tools automatically include the actual timing delays for the synthesized logic circuit. A CPLD timing model supplied by the CAD tool vendor is used to automatically generate the physical device time delays inside the CPLD. Sometimes this timing model is also written in VHDL. For a quick overview of VHDL, several constructs that can be used to synthesize common digital hardware devices will be presented.

6.4 VHDL Synthesis Models of Gate Networks

The first example consists of a simple gate network. In this model, both a concurrent assignment statement and a sequential process are shown which generate the same gate network. X is the output on one network and Y is the output on the other gate network. The two gate networks operate in parallel.

In VHDL synthesis, inputs and outputs from the port declaration in the module will become I/O pins on the programmable logic device. Comment lines begin with "—". The MAX+PLUS II editor performs syntax coloring and is useful to quickly find major problems with VHDL syntax.

Inside a process, statements are executed in sequential order, and all processes are executed in parallel. If multiple assignments are made to a signal inside a process, the last assignment is taken as the new signal value.

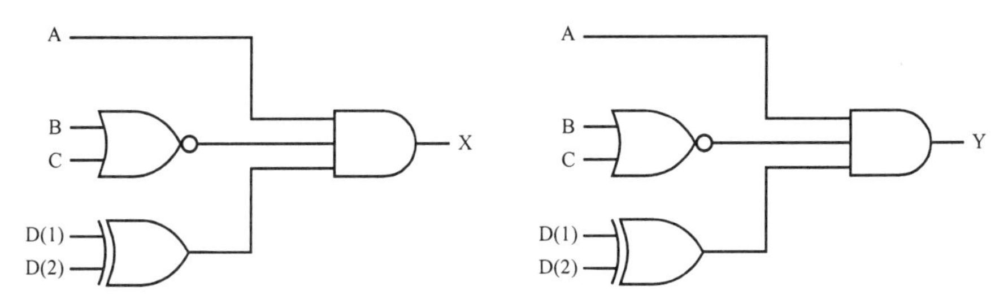


```
LIBRARY IEEE;                          -- Include Libraries for standard logic data types
USE  IEEE.STD_LOGIC_1164.ALL;
                                       -- Entity name normally the same as file name
ENTITY gate_network IS                 -- Ports: Declares module inputs and outputs
        PORT( A, B, C : IN     STD_LOGIC;
                                       -- Standard Logic Vector ( Array of 4 Bits )
              D       : IN     STD_LOGIC_VECTOR( 3 DOWNTO 0 );
                                       -- Output Signals
              X, Y    : OUT    STD_LOGIC );
END gate_network;

                                  -- Defines internal module architecture
ARCHITECTURE behavior OF gate_network IS
BEGIN                             -- Concurrent assignment statements operate in parallel
                                  -- D(1) selects bit 1 of standard logic vector D
        X <= A AND NOT( B OR C ) AND ( D( 1 ) XOR D( 2 ) );

                                  -- Process must declare a sensitivity list,
                                  -- In this case it is  ( A, B, C, D )
                                  -- List includes all signals that can change the outputs
   PROCESS ( A, B, C, D )
        BEGIN                     -- Statements inside process execute sequentially
              Y <= A AND NOT( B OR C) AND ( D( 1) XOR D( 2 ) );
   END PROCESS;
END behavior;
```

6.5 VHDL Synthesis Model of a Seven-segment LED Decoder

The following VHDL code implements a seven-segment decoder for the UP 1's seven-segment LED displays. The displays were described in section 2.3. A 7-bit standard logic vector is used to assign the value of all seven bits in a single case statement. In the logic vector, the most-significant bit is segment 'a' and the least-significant bit is segment 'g'. The logic synthesis CAD tool automatically minimizes the logic required for implementation. The signal MSD contains the 4-bit binary value to be displayed in hexadecimal. MSD is the left or most-significant digit. Another identical process with a different input variable is needed for the second display digit.

```
LED_MSD_DISPLAY:                -- BCD to 7 Segment Decoder for LED Displays

PROCESS (MSD)
BEGIN
                                -- Case statement implements a logic truth table
  CASE MSD IS
      WHEN "0000" =>
                MSD_7SEG <= "1111110";
      WHEN "0001" =>
                MSD_7SEG <= "0110000";
      WHEN "0010" =>
                MSD_7SEG <= "1101101";
      WHEN "0011" =>
                MSD_7SEG <= "1111001";
      WHEN "0100" =>
                MSD_7SEG <= "0110011";
      WHEN "0101" =>
                MSD_7SEG <= "1011011";
      WHEN "0110" =>
                MSD_7SEG <= "1011111";
      WHEN "0111" =>
                MSD_7SEG <= "1110000";
      WHEN "1000" =>
                MSD_7SEG <= "1111111";
      WHEN "1001" =>
                MSD_7SEG <= "1111011";
      WHEN OTHERS =>
                MSD_7SEG <= "0111110";
  END CASE;

END PROCESS LED_MSD_DISPLAY;
```


The following VHDL concurrent assignment statements provide the value to be displayed and connect the individual segments. NOT is used since a logic zero actually turns on the LED. Automatic minimization in the synthesis process will eliminate the extra inverter in the logic circuit. Pin assignments for the seven-segment display must be included in the project's *.acf file or in the top-level schematic.

```
                                -- Provide 4-bit value to display
MSD <= PC ( 7 DOWNTO 4 );
                                -- Drive the seven-segments (LEDs are active low)
MSD_a <= NOT MSD_7SEG( 6 );
MSD_b <= NOT MSD_7SEG( 5 );
MSD_c <= NOT MSD_7SEG( 4 );
MSD_d <= NOT MSD_7SEG( 3 );
MSD_e <= NOT MSD_7SEG( 2 );
MSD_f  <= NOT MSD_7SEG( 1 );
MSD_g <= NOT MSD_7SEG( 0 );
```

6.6 VHDL Synthesis Model of a Multiplexer

The next example shows several alternative ways to synthesize a 2-to-1 multiplexer in VHDL. Four identical multiplexers that operate in parallel are synthesized by this example. In VHDL, IF and CASE statements must be inside a process. The inputs and outputs from the multiplexers could be changed to standard logic vectors if an entire bus is multiplexed. Multiplexers with more than two inputs can also be easily constructed. Nested IF-THEN-ELSE statements generate priority-encoded logic that requires more hardware and produce a slower circuit than a CASE statement.

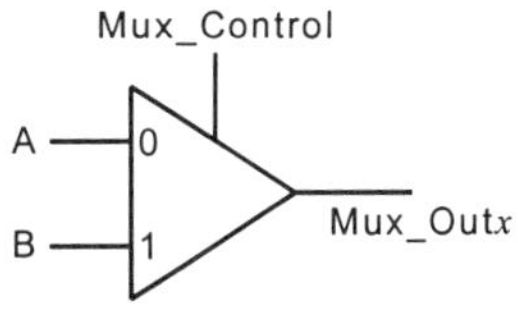

```
LIBRARY IEEE;
USE  IEEE.STD_LOGIC_1164.ALL;

ENTITY multiplexer IS                              -- Input Signals and Mux Control
        PORT( A, B, Mux_Control          : IN     STD_LOGIC;
              Mux_Out1, Mux_Out2,
              Mux_Out3, Mux_Out4    : OUT   STD_LOGIC );
END multiplexer;

ARCHITECTURE behavior OF multiplexer IS
BEGIN                                               -- selected signal assignment statement...

        Mux_Out1 <= A WHEN Mux_Control = '0' ELSE B;
                                                    -- ... with Select Statement
        WITH Mux_control SELECT

        Mux_Out2 <=   A WHEN    '0',
                      B WHEN    '1',
                      A WHEN OTHERS;               -- OTHERS case required since STD_LOGIC
                                                    --     has values other than "0" or "1"
        PROCESS ( A, B, Mux_Control)
        BEGIN                                       -- Statements inside a process
                IF Mux_Control = '0' THEN           --     execute sequentially.
                  Mux_Out3 <= A;
                ELSE
                  Mux_out3 <= B;
                END IF;

                CASE Mux_Control IS
                        WHEN '0' =>
                                Mux_Out4 <= A;
                        WHEN '1' =>
                                Mux_Out4 <= B;
                        WHEN OTHERS =>
                                Mux_Out4 <= A;
                END CASE;
        END PROCESS;
END behavior;
```

6.7 VHDL Synthesis Model of Tri-State Output

Tri-state gates are supported in VHDL synthesis tools and are supported in many programmable logic devices. Most programmable logic devices have tri-state output pins. Some programmable logic devices do not support internal tri-state logic. Here is a VHDL example of a tri-state output. In VHDL, the assignment of the value "Z" to a signal produces a tri-state output.

Control
A
Tri_Out

```
LIBRARY IEEE;
USE  IEEE.STD_LOGIC_1164.ALL;

ENTITY tristate IS
   PORT( A, Control      : IN      STD_LOGIC;
            Tri_out      : INOUT   STD_LOGIC); -- Use Inout for bi-directional tri-state
                                                --    signals or out for output only
END tristate;

ARCHITECTURE behavior OF tristate IS               -- defines internal module architecture
BEGIN
   Tri_out <= A WHEN Control = '0' ELSE 'Z';       -- Assignment of 'Z' value generates
END behavior;                                       --    tri-state output
```

6.8 VHDL Synthesis Models of Flip-flops and Registers

In the next example, several flip-flops will be generated. Unlike earlier combinational hardware devices, a flip-flop can only be synthesized inside a process. In VHDL, Clock'EVENT is true whenever the clock signal changes. The positive clock edge is selected by (clock'EVENT AND clock = '1') and positive edge triggered D flip-flops will be used for synthesis. The following module contains a variety of Reset and Enable options on positive edge-triggered D flip-flops. Processes with a wait statement do not need a process sensitivity list. A process can only have one clock or reset type.

The negative clock edge is selected by (clock'EVENT AND clock = '0') and negative edge-triggered D flip-flops will be used for synthesis. If (Clock = '1') is substituted for (clock'EVENT AND clock = '1') level-triggered latches will be selected for logic synthesis. Rising_edge(clock) can also be used instead of clock'EVENT AND clock = '1'.

```
LIBRARY IEEE;
USE  IEEE.STD_LOGIC_1164.ALL;

ENTITY DFFs IS
    PORT( D, Clock, Reset, Enable  : IN    STD_LOGIC;
           Q1, Q2, Q3, Q4          : OUT  STD_LOGIC );
END DFFs;

ARCHITECTURE behavior OF DFFs IS
BEGIN
```

```
    PROCESS                                  -- Positive edge triggered D flip-flop
    BEGIN                                    -- If WAIT is used no sensitivity list is used
        WAIT UNTIL ( Clock 'EVENT AND Clock = '1' );
            Q1 <= D;
    END PROCESS;

    PROCESS                                  -- Positive edge triggered D flip-flop
    BEGIN                                    --     with synchronous reset
        WAIT UNTIL ( Clock 'EVENT AND Clock = '1' );
            IF reset = '1' THEN
                Q2 <= '0';
            ELSE
                Q2 <= D;
            END IF;
    END PROCESS;

    PROCESS (Reset,Clock)                    -- Positive edge triggered D flip-flop
    BEGIN                                    --     with asynchronous reset
        IF reset = '1' THEN
            Q3 <= '0';
        ELSIF ( clock 'EVENT AND clock = '1' ) THEN
            Q3 <= D;
        END IF;
    END PROCESS;

    PROCESS (Reset,Clock)                    -- Positive edge triggered D flip-flop
    BEGIN                                    --     with asynchronous reset and
                                             --     enable
        IF reset = '1' THEN
            Q4 <= '0';
        ELSIF ( clock 'EVENT AND clock = '1' ) THEN
            IF Enable = '1' THEN
                Q4 <= D;
            END IF;
        END IF;
    END PROCESS;
END behavior;
```

In VHDL, as in any digital logic designs, it is not good practice to AND or gate other signals with the clock. Use a flip-flop with a clock enable instead to avoid timing and clock skew problems. In some limited cases, such as power management, a single level of clock gating can be used. This works only when a small amount of clock skew can be tolerated and the signal gated with the clock is known to be hazard or glitch free. A particular programmable logic device may not support every flip-flop or latch type and Set/Reset and Enable option.

If D and Q are replaced by standard logic vectors in these examples, registers with the correct number of bits will be generated instead of individual flip-flops.

6.9 Accidental Synthesis of Inferred Latches

Here is a very common problem to be aware of when coding VHDL for synthesis. If a non-clocked process has any path that does not assign a value to an output, VHDL assumes you want to use the previous value. A level triggered latch is automatically generated or inferred by the synthesis tool to save the previous value. In many cases, this can cause serious errors in the design. Edge-triggered flip-flops should not be mixed with level-triggered latches in a design or serious timing problems will result. Typically this can happen in CASE statements or nested IF statements. In the following example, the signal OUTPUT2 infers a latch when synthesized. Assigning a value to OUTPUT2 in the last ELSE clause will eliminate the latch.

```
LIBRARY IEEE;
USE  IEEE.STD_LOGIC_1164.ALL;
ENTITY ilatch IS
        PORT( A, B                      : IN      STD_LOGIC;
                Output1, Output2        : OUT     STD_LOGIC );
END ilatch;

ARCHITECTURE behavior OF ilatch IS
BEGIN
        PROCESS ( A, B )
        BEGIN
                IF A = '0' THEN
                    Output1 <= '0';
                    Output2 <= '0';
                ELSE
                    IF B = '1' THEN
                        Output1 <= '1';
                        Output2 <= '1';
                    ELSE                    -- Latch inferred since no value is assigned
                        Output1 <= '0';     --    to output2 in the else clause!
                    END IF;
                END IF;
        END PROCESS;
END behavior;
```

6.10 VHDL Synthesis Model of a Counter

Here is an 8-bit counter design. This design performs arithmetic operations on standard logic vectors. Since this example includes arithmetic operations, two new libraries must be included at the beginning of the module. Either signed or unsigned libraries can be selected, but not both. Since the unsigned library was used, an 8-bit magnitude comparator is automatically synthesized for the internal_count < max_count comparison.

Compare operations between standard logic and integer types are supported. The assignment internal_count <= internal_count + 1 synthesizes an 8-bit incrementer. An incrementer circuit requires less hardware than an adder that adds one. The operation, "+1", is treated as a special incrementer case by synthesis tools. VHDL does not allow reading of an "OUT" signal so an internal_count signal is used which is always the same as count. This is the first example that includes an internal signal. Note its declaration at the beginning of the architecture section.

```
LIBRARY IEEE;
USE IEEE.STD_LOGIC_1164.ALL;
USE IEEE.STD_LOGIC_ARITH.ALL;
USE IEEE.STD_LOGIC_UNSIGNED.ALL;

ENTITY Counter IS
    PORT( Clock, Reset   : IN   STD_LOGIC;
          Max_count      : IN   STD_LOGIC_VECTOR( 7 DOWNTO 0 );
          Count          : OUT  STD_LOGIC_VECTOR( 7 DOWNTO 0 ) );
END Counter;

ARCHITECTURE behavior OF Counter IS                  -- Declare signal(s) internal to module
    SIGNAL internal_count:      STD_LOGIC_VECTOR( 7 DOWNTO 0 );
BEGIN
    count <= internal_count;

    PROCESS ( Reset,Clock )
        BEGIN                                          -- Reset counter
            IF reset = '1' THEN
                internal_count <= "00000000";
            ELSIF ( clock 'EVENT AND clock = '1' ) THEN
                IF internal_count < Max_count THEN     -- Check for maximum count
                    internal_count <= internal_count + 1;  -- Increment Counter
                ELSE                                   -- Count >= Max_Count
                    internal_count <= "00000000";      --    reset Counter
                END IF;
            END IF;
    END PROCESS;
END behavior;
```

6.11 VHDL Synthesis Model of a State Machine

The next example shows a Moore state machine with three states, two inputs and a single output. A state diagram of the example state machine is shown in Figure 6.1. In VHDL, an enumerated data type is specified for the current state using the TYPE statement. This allows the synthesis tool to assign the actual "0" or "1" values to the states. In many cases, this will produce a smaller hardware design than direct assignment of the state values in VHDL.

Depending on the synthesis tool settings, the states may be encoded or constructed using the one-hot technique. Outputs are defined in the last WITH... SELECT statement. This statement lists the output for each state and eliminates possible problems with inferred latches. To avoid possible timing problems, unsynchronized external inputs to a state machine should be synchronized by passing them through one or two D flip-flops that are clocked by the state machine's clock.

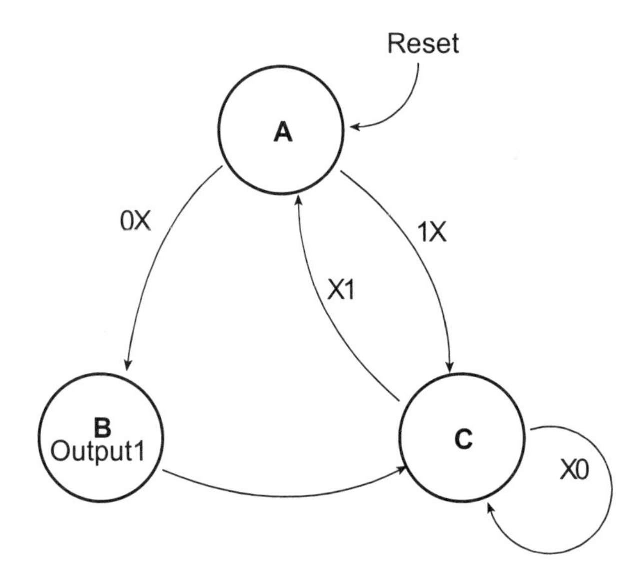

Figure 6.1 State Diagram for st_mach VHDL example

```
LIBRARY IEEE;
USE IEEE.STD_LOGIC_1164.ALL;

ENTITY st_mach IS
    PORT( clk, reset         : IN     STD_LOGIC;
          Input1, Input2     : IN     STD_LOGIC;
          Output1            : OUT    STD_LOGIC);
END st_mach;

ARCHITECTURE A OF st_mach IS
                                        -- Enumerated Data Type for State
    TYPE STATE_TYPE IS ( state_A, state_B, state_C );
    SIGNAL state: STATE_TYPE;

BEGIN
    PROCESS ( reset, clk )
    BEGIN
        IF reset = '1' THEN             -- Reset State
            state <= state_A;
        ELSIF clk 'EVENT AND clk = '1' THEN

            CASE state IS               -- Define Next State Transitions using a Case
                                        --     Statement based on the Current State
                WHEN state_A =>
```

```
                IF Input1 = '0' THEN
                    state <= state_B;
                ELSE
                    state <= state_C;
                END IF;

            WHEN state_B =>
                state <= state_C;

            WHEN state_C =>
                IF Input2 = '1' THEN
                    state <= state_A;
                END IF;

            WHEN OTHERS =>
                state <= state_A;
        END CASE;
    END IF;
  END PROCESS;

  WITH state SELECT                          -- Define State Machine Outputs
    Output1  <=   '0' WHEN state_A,
                  '1' WHEN state_B,
                  '0' WHEN state_C;

END a;
```

6.12 VHDL Synthesis Model of an ALU with an Adder/Subtractor and a Shifter

Here is an 8-bit arithmetic logic unit (ALU), that adds, subtracts, bitwise ANDs, or bitwise ORs, two operands and then performs an optional shift on the output. The most-significant two bits of the Op-code select the arithmetic logical operation. If the least-significant bit of the op_code equals '1' a 1-bit left-shift operation is performed. An addition and subtraction circuit is synthesized for the "+" and "-" operator. Depending on the number of bits and the speed versus area settings in the synthesis tool, ripple carry or carry-lookahead circuits will be used. Several "+" and "-" operations in multiple assignment statements may generate multiple ALUs and increase the hardware size, depending on the VHDL CAD tool and compiler settings used. If a single ALU is desired, muxes can be placed at the inputs and the "+" operator would be used only in a single assignment statement.

```
LIBRARY IEEE;
USE  IEEE.STD_LOGIC_1164.ALL;
USE  IEEE.STD_LOGIC_ARITH.ALL;
USE  IEEE.STD_LOGIC_UNSIGNED.ALL;

ENTITY ALU IS
   PORT( Op_code                  : IN      STD_LOGIC_VECTOR( 2 DOWNTO 0 );
```

```
            A_input, B_input        : IN    STD_LOGIC_VECTOR( 7 DOWNTO 0 );
            ALU_output              : OUT   STD_LOGIC_VECTOR( 7 DOWNTO 0 ) );
END ALU;

ARCHITECTURE behavior OF ALU IS
                                          -- Declare signal(s) internal to module here
    SIGNAL temp_output              :     STD_LOGIC_VECTOR( 7 DOWNTO 0 );
BEGIN

    PROCESS ( Op_code, A_input, B_input )
    BEGIN
        CASE Op_Code ( 2 DOWNTO 1 ) IS    -- Select Arithmetic/Logical Operation
            WHEN "00" =>
                    temp_output <= A_input  +   B_input;
            WHEN "01" =>
                    temp_output <= A_input  -   B_input;
            WHEN "10" =>
                    temp_output <= A_input  AND B_input;
            WHEN "11" =>
                    temp_output <= A_input  OR  B_input;
            WHEN OTHERS =>
                    temp_output <= "00000000";
        END CASE;

        -- Select Shift Operation: Shift bits left with zero fill using concatenation operator
        --   Can also use VHDL 1076-1993 shift operator such as SLL

        IF Op_Code( 0 ) = '1' THEN
            Alu_output <= temp_output( 6 DOWNTO 0 ) & '0';
        ELSE
            Alu_output <= temp_output;
        END IF;
    END PROCESS;
END behavior;
```

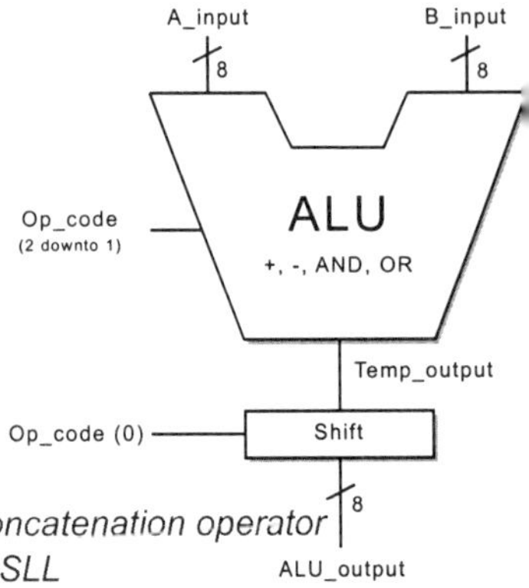

6.13 VHDL Synthesis of Multiply and Divide Hardware

In the MAX+PLUS II tool, only multiply and divide by powers of two (shifts) are supported using VHDL's "*" and "/" operators. Mod and Rem are not supported in MAX+PLUS II. In current generation tools, efficient design of multiply or divide hardware typically requires the use of a vendor-specific library function or even the specification of the arithmetic algorithm and hardware implementation in VHDL.

A wide variety of multiply and divide algorithms that trade off time versus hardware size can be found in most computer arithmetic texts. Several such references are listed at the end of this chapter. These algorithms require a sequence of add/subtract and shift operations that can be easily synthesized in VHDL using the standard operators. The LPM_MULT function in MAX+PLUS II can be used to synthesize integer multipliers. LPM_DIVIDE, is also available. When using LPM functions, **File ⇨Megawizard** can be used to help generate VHDL code. The LPM functions also support pipeline options. Array multiply and divide hardware for more than a few bits requires extensive hardware and a large FPLD.

```
LIBRARY IEEE;
USE IEEE.STD_LOGIC_1164.ALL;
USE IEEE.STD_LOGIC_ARITH.ALL;
USE IEEE.STD_LOGIC_UNSIGNED.ALL;
LIBRARY lpm;
USE lpm.lpm_components.ALL;

ENTITY mult IS
      PORT( A, B      : IN   STD_LOGIC_VECTOR( 7 DOWNTO 0 );
            Product   : OUT STD_LOGIC_VECTOR( 15 DOWNTO 0 ) );
END mult;

ARCHITECTURE a OF mult IS
BEGIN                                   -- LPM 8x8 multiply function P = A * B
      multiply: lpm_mult
      GENERIC MAP( LPM_WIDTHA              => 8,
                   LPM_WIDTHB              => 8,
                   LPM_WIDTHS              => 16,
                   LPM_WIDTHP              => 16,
                   LPM_REPRESENTATION      => "UNSIGNED" )

      PORT MAP (   data  => A,
                   datab => B,
                   result => Product );
END a;
```

Floating-point operations can be implemented on very large FPLDs; however, performance is lower than current floating-point DSP and microprocessor chips. The floating-point algorithms must be coded by the user in VHDL using integer add, multiply, divide, and shift operations. The LPM_CLSHIFT function is useful for the barrel shifter needed in a floating-point ALU. Many FPLD vendors also have optimized arithmetic packages for DSP applications such as FIR filters.

6.14 VHDL Synthesis Models for Memory

Typically, it is more efficient to call a vendor-specific function to synthesize RAM. The memory function in the Altera toolset is the LPM_RAM_DQ function. The memory can be set to an initial value using a separate memory initialization file with the extension *.mif. A similar call, LPM_ROM, can be used to synthesize ROM.

If small blocks of multi-ported or other special-purpose RAM are needed, they can be synthesized using registers with address decoders for the write operation and multiplexers for the read operation. Additional read or write ports can be added to synthesized RAM. An example of this approach is a dual-ported register file for a computer processor core. Most RISC processors need to read two registers on each clock cycle and write to a third register.

VHDL Memory Model - Example One

The first memory example synthesizes a memory that can perform a read and a write operation every clock cycle. Memory is built using arrays of positive edge-triggered D flip-flops. Memory write, memwrite, is gated with an address decoder output and used as an enable to load each memory location during a write operation. A synchronous write operation is more reliable. Asynchronous write operations respond to any logic hazards or momentary level changes on the write signal. As in any synchronous memory, the write address must be stable before the rising edge of the clock signal. A non-clocked mux is used for the read operation. If desired, memory can be initialized by a reset signal.

```
LIBRARY IEEE;
USE IEEE.STD_LOGIC_1164.ALL;

ENTITY memory IS
     PORT( read_data       :     OUT  STD_LOGIC_VECTOR( 7 DOWNTO 0 );
            read_address   :     IN   STD_LOGIC_VECTOR( 2 DOWNTO 0 );
            write_data     :     IN   STD_LOGIC_VECTOR( 7 DOWNTO 0 );
            write_address  :     IN   STD_LOGIC_VECTOR( 2 DOWNTO 0 );
            Memwrite       :     IN   STD_LOGIC;
            clock,reset    :     IN   STD_LOGIC  );
END memory;

ARCHITECTURE behavior OF memory IS
     SIGNAL mem0, mem1  :             STD_LOGIC_VECTOR( 7 DOWNTO 0 );

BEGIN

     PROCESS (read_address, mem0, mem1)  -- Process for memory read operation
     BEGIN
          CASE read_address IS
               WHEN "000" =>
                         read_data <= mem0;
               WHEN "001" =>
                         read_data <= mem1;
               WHEN OTHERS =>                 -- Unimplemented memory locations
                         read_data <= To_stdlogicvector( X"FF" );
          END CASE;
     END PROCESS;

     PROCESS
     BEGIN
          WAIT UNTIL clock 'EVENT AND clock = '1';
               IF ( reset = '1' ) THEN
                    mem0 <= To_stdlogicvector( X"55" ); -- Initial values for memory (optional)
                    mem1 <= To_stdlogicvector( X"AA" );
               ELSE
                    IF memwrite = '1' THEN              -- Write to memory?
                    CASE write_address IS               -- Use a flip-flop with
```

```
                        WHEN "000" =>              --       an enable for memory
                            mem0 <= write_data;
                        WHEN "001" =>
                            mem1 <= write_data;
                        WHEN OTHERS =>             -- unimplemented memory locations
                            NULL;
                END CASE;
                END IF;
            END IF;
    END PROCESS;
END behavior;
```

VHDL Memory Model - Example Two

The second example uses an array of standard logic vectors to implement memory. This approach is easier to write in VHDL since the array index generates the address decoder and multiplexers automatically; however, it is a little more difficult to access the values of individual array elements during simulation. There are a few VHDL synthesis tools that do not support array types. Synthesizing RAM requires a vast amount of programmable logic resources. Only a few hundred bits of RAM can be synthesized, even on large devices. Each bit of RAM requires 10 to 20 logic gates and a large amount of FPLD interconnect resources.

```
LIBRARY IEEE;
USE IEEE.STD_LOGIC_1164.ALL;

ENTITY memory IS
    PORT( read_data     :   OUT  STD_LOGIC_VECTOR( 7 DOWNTO 0 );
          read_address  :   IN   STD_LOGIC_VECTOR( 2 DOWNTO 0 );
          write_data    :   IN   STD_LOGIC_VECTOR( 7 DOWNTO 0 );
          write_address :   IN   STD_LOGIC_VECTOR( 2 DOWNTO 0 );
          Memwrite      :   IN   STD_LOGIC;
          Clock         :   IN   STD_LOGIC  );
END memory;
ARCHITECTURE behavior OF memory IS
                                        -- define new data type for memory array
    TYPE memory_type IS ARRAY ( 0 TO 7 ) OF STD_LOGIC_VECTOR( 7 DOWNTO 0 );
    SIGNAL memory        : memory_type;
BEGIN
    -- Read Memory and  convert array index to an integer with CONV_INTEGER
    read_data <= memory( CONV_INTEGER( read_address( 2 DOWNTO 0 ) ) );

    PROCESS                        -- Write Memory?
    BEGIN
        WAIT UNTIL clock 'EVENT AND clock = '1';
        IF ( memwrite = '1' ) THEN
                                   -- convert array index to an integer with CONV_INTEGER
          memory( CONV_INTEGER( write_address( 2 DOWNTO 0 ) ) ) <= write_data;
        END IF;
    END PROCESS;
END behavior;
```

VHDL Memory Model - Example Three

The third example shows the use of the LPM_RAM_DQ megafunction to implement a block of memory. An additional library is needed for the LPM functions. For more information on the LPM functions see the online help guide in the MAX+PLUS II tool. The LPM_RAM memory can do either a read or a write operation in a single clock cycle since there is only one address bus. If this is the only memory operation needed, the LPM_RAM function produces a more efficient hardware implementation than synthesis of the memory in VHDL. In this LPM_RAM_DQ megafunction, the memory address must be stable whenever write enable is High. On the FLEX chip this function can be implemented using the EAB memory blocks, which are separate from the logic cells. In the FLEX 10K20 chip, six 2K-bit EABs provide a total of 12K bits of available memory. The FLEX 10K70 on the newer UP-1X board contains nine 2K-bit EABs.

```
LIBRARY IEEE;
USE IEEE.STD_LOGIC_1164.ALL;

ENTITY amemory IS
    PORT( read_data         :   OUT   STD_LOGIC_VECTOR( 7 DOWNTO 0 );
          memory_address    :   IN    STD_LOGIC_VECTOR( 2 DOWNTO 0 );
          write_data        :   IN    STD_LOGIC_VECTOR( 7 DOWNTO 0 );
          Memwrite          :   IN    STD_LOGIC;
          clock,reset       :   IN    STD_LOGIC );
END amemory;

ARCHITECTURE behavior OF amemory IS
BEGIN
    data_memory: lpm_ram_dq             -- LPM memory function

    GENERIC MAP ( lpm_widthad            => 3,
                  lpm_outdata            => "UNREGISTERED",
                  lpm_indata             => "REGISTERED",
                  lpm_address_control    => "UNREGISTERED",
                                      -- Reads in mif file for initial data values (optional)
                  lpm_file               => "memory.mif",
                  lpm_width              => 8 )

    PORT MAP ( data => write_data, address  => memory_address( 2 DOWNTO 0 ),
               We   => Memwrite, inclock   => clock, q => read_data );
END behavior;
```

6.15 Hierarchy in VHDL Synthesis Models

Large VHDL models should be split into a hierarchy using a top-level structural model in VHDL or by using the symbol and graphic editor in the MAX+PLUS II tool. In the graphical editor, a VHDL file can be used to define the contents of a symbol block. Synthesis tools run faster using a hierarchy on large models and it is easier to write, understand, and maintain a large design when it is broken up into smaller modules.

An example of a hierarchical design with three submodules is seen in the schematic in Figure 6.2. Following the schematic, the same design using a top-level VHDL structural model is shown. This VHDL structural model provides the same connection information as the schematic seen in Figure 6.2.

Debounce, Onepulse, and Clk_div are the names of the VHDL submodules. Each one of these submodules has a separate VHDL source file. In the MAX+PLUS II tool, compiling the top-level module will automatically compile the lower-level modules.

In the example, VHDL structural model, note the use of a component declaration for each submodule. The component statement declares the module name and the inputs and outputs of the module. Internal signal names used for interconnections of components must also be declared at the beginning of the component list.

In the final section, port mappings are used to specify the module or component interconnections. Port names and their order must be the same in the VHDL submodule file, the component instantiations, and the port mappings. Component instantiations are given unique labels so that a single component can be used several times.

Note that node names in the schematic or signals in VHDL used to interconnect modules need not always have the same names as the signals in the components they connect. As an example, pb_debounced on the debounce component connects to an internal signal with a different name, pb1_debounced.

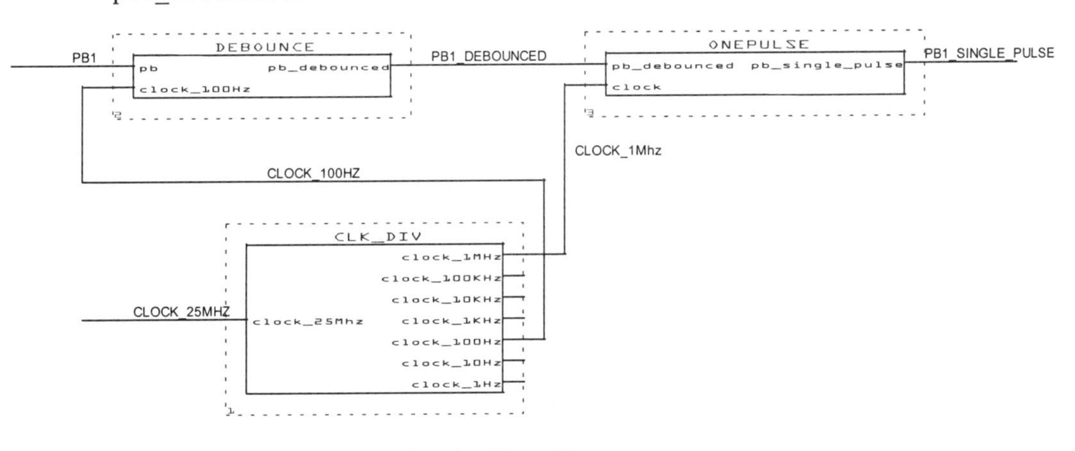

Figure 6.2 Schematic of Hierarchical Design Example

```
LIBRARY IEEE;
USE IEEE.STD_LOGIC_1164.ALL;
USE IEEE.STD_LOGIC_ARITH.ALL;
USE IEEE.STD_LOGIC_UNSIGNED.ALL;
ENTITY hierarch IS
    PORT (  clock_25Mhz, pb1     : IN STD_LOGIC;
            pb1_single_pulse     : OUT STD_LOGIC);
```

```
END hierarch;
ARCHITECTURE structural OF hierarch IS
-- Declare internal signals needed to connect submodules
SIGNAL clock_1MHz, clock_100Hz, pb1_debounced : STD_LOGIC;
-- Use Components to Define Submodules and Parameters
   COMPONENT debounce
           PORT(    pb, clock_100Hz        : IN      STD_LOGIC;
                    pb_debounced           : OUT     STD_LOGIC);
   END COMPONENT;

   COMPONENT onepulse
           PORT(pb_debounced, clock        : IN      STD_LOGIC;
            pb_single_pulse                : OUT     STD_LOGIC);
   END COMPONENT;

   COMPONENT clk_div
           PORT(    clock_25Mhz            : IN      STD_LOGIC;
                    clock_1MHz             : OUT     STD_LOGIC;
                    clock_100KHz           : OUT     STD_LOGIC;
                    clock_10KHz            : OUT     STD_LOGIC;
                    clock_1KHz             : OUT     STD_LOGIC;
                    clock_100Hz            : OUT     STD_LOGIC;
                    clock_10Hz             : OUT     STD_LOGIC;
                    clock_1Hz              : OUT     STD_LOGIC);
   END COMPONENT;
BEGIN

-- Use Port Map to connect signals between components in the hierarchy
debounce1 : debounce PORT MAP  (pb => pb1, clock_100Hz = >clock_100Hz,
                          pb_debounced = >pb1_debounced);

prescalar : clk_div  PORT MAP (clock_25Mhz = >clock_25Mhz,
                          clock_1MHz =>clock_1Mhz,
                    clock_100hz = >clock_100hz);

single_pulse : onepulse PORT MAP (pb_debounced = >pb1_debounced,
                          clock => clock_1MHz,
                          pb_single_pulse => pb1_single_pulse);
END structural;
```

6.16 Using a Testbench for Verification

Complex VHDL synthesis models are frequently verified by simulation of the model's behavior in a specially written entity called a testbench. As seen in Figure 6.3, the top–level testbench module contains a component instantiation of the hardware unit under test (UUT). The testbench also contains VHDL code used to automatically generate input stimulus to the UUT and automatically monitor the response of the UUT for correct operation.

The testbench contains test vectors and timing information used in testing the UUT. The testbench's VHDL code is used only for testing, and it is not synthesized. This keeps the test-only code portion of the VHDL model separate from the UUT's hardware synthesis model. In large designs, the testbench can require as much time and effort as the UUT's synthesis model. By performing both a functional simulation and a timing simulation of the UUT with the same test vectors, it is also possible to check for any synthesis-related errors.

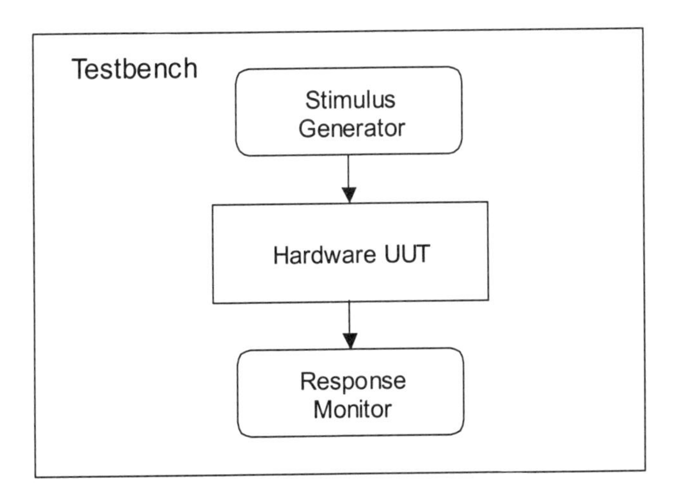

Figure 6.3 Using a testbench for automatic verification during simulation.

6.17 For additional information

The chapter has introduced the basics of using VHDL for digital synthesis. It has not explored all of the language options available. The Altera online help contains VHDL syntax and templates. A large number of VHDL reference textbooks are also available. Unfortunately, only a few of them currently examine using VHDL models that are targeted for digital logic synthesis. One such text is *HDL Chip Design* by Douglas J. Smith, Doone Publications, 1996.

A number of alternative integer multiply, divide, and floating-point algorithms with different speed versus area tradeoffs can be found in computer arithmetic textbooks. Two such examples are *Digital Computer Arithmetic Design and Implementation* by Cavanagh, McGraw Hill, 1984, and *Computer Arithmetic Algorithms* by Israel Koren, Prentice Hall, 1993.

6.18 Laboratory Exercises

1. Rewrite the VHDL model for the seven-segment decoder in Section 6.5. Replace the PROCESS and CASE statements with a WITH...SELECT statement.

2. Write a VHDL model for the state machine shown in the following state diagram and verify correct operation with a simulation using the Altera CAD tools. A and B are the two states, X is the output, and Y is the input. Use the timing analyzer to determine the maximum clock frequency on the FLEX device.

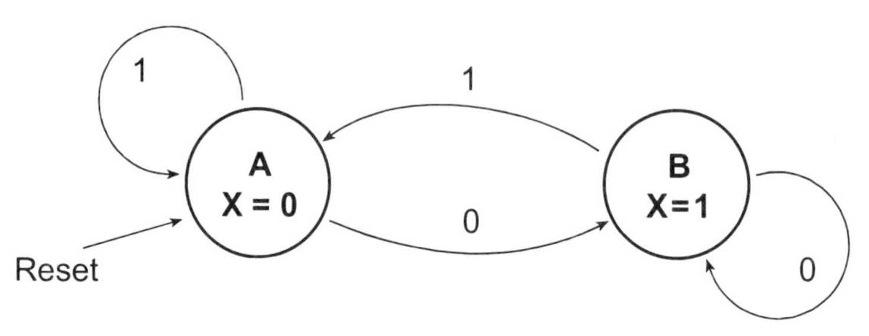

3. Write a VHDL model for a 12-bit, arithmetic logic unit (ALU). Verify correct operation with a simulation using the Altera CAD tools. A and B are 12-bit inputs to the ALU, and Y is the output. A shift operation follows the arithmetic and logical operation. The opcode controls ALU functions as follows:

Opcode	Operation	Function
000XX	ALU_OUT <= A	Pass A
001XX	ALU_OUT <= A + B	Add
010XX	ALU_OUT <= A-B	Subtract
011XX	ALU_OUT <= A AND B	Logical AND
100XX	ALU_OUT <= A OR B	Logical OR
101XX	ALU_OUT <= A + 1	Increment A
110XX	ALU_OUT <= A-1	Decrement A
111XX	ALU_OUT <= B	Pass B
XXX00	Y <= ALU_OUT	Pass ALU_OUT
XXX01	Y<= SHL(ALU_OUT)	Shift Left
XXX10	Y<=SHR(ALU_OUT)	Shift Right (unsigned-zero fill)
XXX11	Y<= 0	Pass 0's

Use the FLEX chip as the target device, because this design is too large for the MAX device. Determine the worst case time delay of the ALU using the timing analyzer. Examine the report file and find the device utilization. Multiply the logic cell (LC) device utilization percentage by 20,000 to estimate the size of the design in gates.

4. Explore different synthesis options for the ALU from problem 3. Change the area and speed synthesis settings in the compiler under **Assign ⇨ Global Project Logic Synthesis**, rerun the timing analyzer to determine speed, and examine the report file for hardware size estimates. Include as a minimum, data points for the default, optimized for speed, and optimized for area settings. Build a plot showing the speed versus area trade-offs possible in the synthesis tool. Multiply the logic cell (LC) device utilization percentage found in the report file by 20,000 to estimate the size of the designs in gates.

5. Develop a VHDL model of one of the TTL chips listed below. The model should be functionally equivalent, but there will be timing differences. Compare the timing

differences between the VHDL FPLD implementation and the TTL chip. Use a data book or find a data sheet using the World Wide Web.

A. 7400 Quad nand gate

B. 74LS241 Octal buffer with tri-state output

C. 74LS273 Octal D flip-flop with Clear

D. 74163 4-bit binary counter

E. 74LS181 4-bit ALU

6. Replace the 8count block used in the tutorial in Chapter 4, with a new counter module written in VHDL. Simulate the design and download a test program to the UP 1 board.

7. Implement a 512 by 8 RAM using VHDL and the LPM_RAM_DQ function. Do not use registered options. Target the design to the FLEX device. Use the timing analyzer to determine the worst-case read and write access times for the memory.

CHAPTER 7

State Machine Design: The Electric Train Controller

7 State Machine Design: The Electric Train Controller

7.1 The Train Control Problem

The track layout of a small electric train system is shown in Figure 7.1. Two trains, we'll call A and B, run on the tracks, hopefully without colliding. To avoid collisions, the trains require a safety controller that allows trains to move in and out of intersections without mishap.

In Figure 7.1, assume for a moment that Train A is at Switch 3 and moving counterclockwise. Let's also assume that Train B is moving counterclockwise and is at Sensor 2. Since Train B is entering the common track (Track 2), Train A must be stopped when it reaches Sensor 1, and must wait until Train B has passed Sensor 3. At this point, Train A will be allowed to enter Track 2, and Train B will move toward Sensor 2.

The controller is a state machine that uses the sensors as inputs. The controller's outputs control the power to the tracks, the direction of the trains, and the position of the switches. However, the state machine does not control the speed of the train. This means that the system controller must function correctly independent of the speed of the two trains.

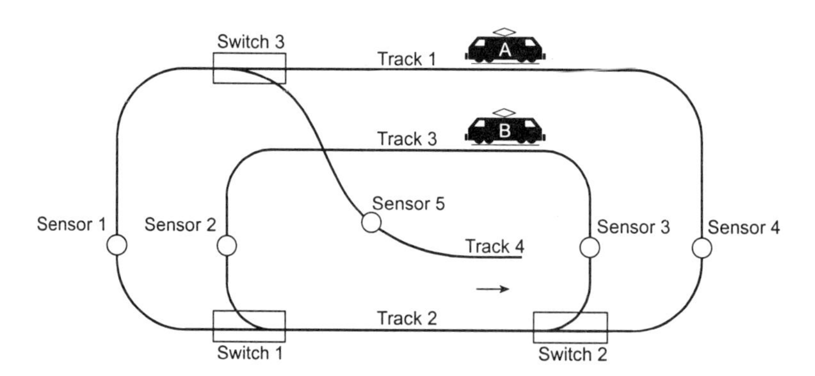

Figure 7.1 Track Layout with Input Sensors and Output Switches and Output Tracks.

The following sections describe how the state machine should control each signal to operate the trains properly. Figure 7.2 demonstrates how the actual electric train system was built with relays. A relay is an electrically controlled switch. A UP 1-based "virtual" train simulation will be used that emulates this relay setup. Since there are no actual relays on the UP 1 board, it is only intended to give you a visual diagram of how the output signals work in the real system.

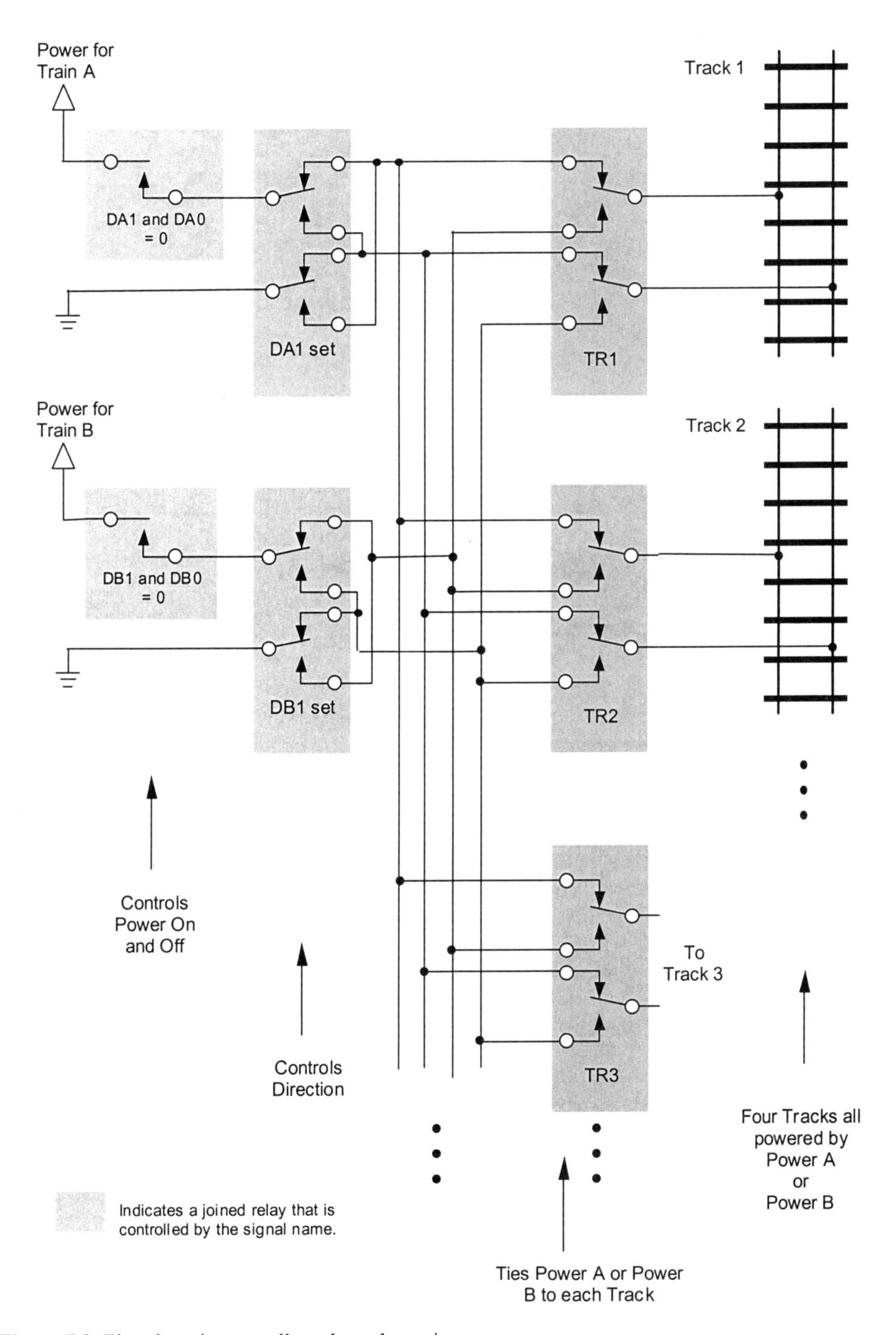

Figure 7.2 Electric train controller relay schematic.

7.2 Track Power (T1, T2, T3, and T4)

The track power signals (T1, T2, T3, T4) determine which power supply is attached to which track. Note that since these are binary signals, one of the power supplies is always connected to a track.

The track power connections are based on the actual switching relay system. There are two power supplies (Power A and Power B) that can be connected to the tracks. What these two power supplies actually allow you to do on the real system is assign different speeds to each train. This is part of the simulator and is controlled by the DIP switches on the UP 1 board. Speed will NOT be controlled by your state machine, only Stop, Forward, and Reverse.

As illustrated in Figure 7.2, if either direction switch is set, the power supply is connected to the next level of switches. Another set of relays determines the polarity (direction of trains), while a third connects either Power A or Power B to each track. Each track is either assigned power from source A (Tn = 0) or from source B (Tn = 1). In other words, if all four signals (T1, T2, T3, and T4) are asserted high, all tracks will be powered from power source B and would all be assigned the same Direction (see the next section under Track Direction for controlling train direction). See Figure 7.3 for an example.

7.3 Track Direction (DA1-DA0, and DB1-DB0)

The direction for each track is controlled by four signals (two for each power source), DA (DA1-DA0) for source A, and DB (DB1-DB0) for source B. When these signals indicate forward "01" for a particular power source, any train on a track assigned to that power source will move counterclockwise (on track 4, the train moves toward the outer track). When the signals imply reverse "10", the train(s) will move clockwise. The "11" value is illegal and should not be used. When these signals are set to "00", any train assigned to the given source will stop. (See Figures 7.2 and 7.3.)

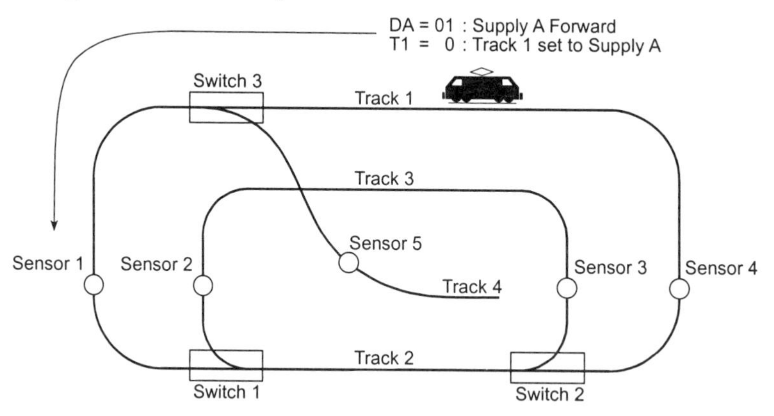

Figure 7.3 Track Power is connected to one of Two Power Sources: A and B.

7.4 Switch Direction (SW1, SW2, and SW3)

Switch directions are controlled by asserting SW1, SW2, and SW3 either high (outside connected with inside track) or low (outside tracks connected). That is, anytime all of the switches are set to 1, the tracks are setup such that the outside tracks are connected to the inside tracks. (See Figure 7.4.)

If a train moves the wrong direction through an open switch it will derail. Be careful. If a train is at the point labeled "Track 1" in Figure 7.4 and is moving to the left, it will derail at Switch 3. To keep it from derailing, SW3 would need to be set to 0.

Also, note that Tracks 3 and 4 cross at an intersection and care must be taken to avoid a crash at this point.

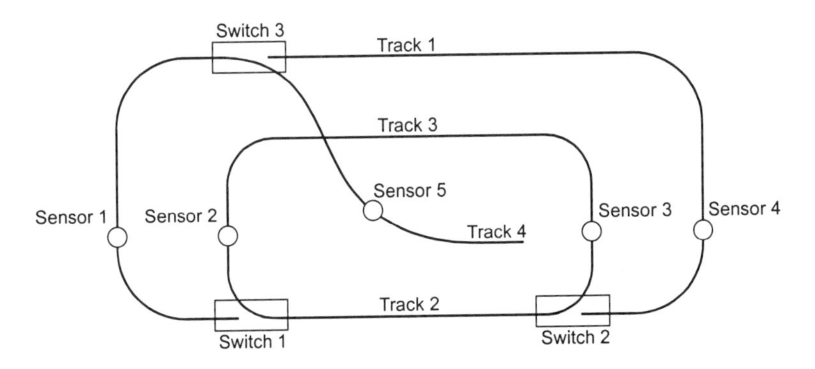

Figure 7.4 Track Direction if all Switches are Asserted (SW1 = SW2 = SW3 = 1)

7.5 Train Sensor Input Signals (S1, S2, S3, S4, and S5)

The five train sensor inputs (S1, S2, S3, S4, and S5) go high when a train is near the sensor location. It should be noted that sensors (S1, S2, S3, S4, and S5) *do not go high for only one clock cycle*. In fact, the sensors fire continuously for *many* clock cycles per passage of a train. This means that if your design is testing the same sensor from one state to another, you must wait for the signal to change from high to low.

As an example, if you wanted to count how many times that a train passes Sensor 1, you can not just have an "IF S1 GOTO count-one state" followed by "IF S1 GOTO count-two state." You would need to have a state that sees S1='1', then S1='0', then S1='1' again before you can be sure that it has passed S1 twice. If your state machine has two concurrent states that look for S1='1', the state machine will pass through both states in two consecutive clock cycles although the train will have passed S1 only once.

Another way would be to detect S1='1', then S4='1', then S1='1' if, in fact, the train was traversing the outside loop continuously. Either method will ensure that the train passed S1 twice.

The signal inputs and outputs have been summarized in the following figure:

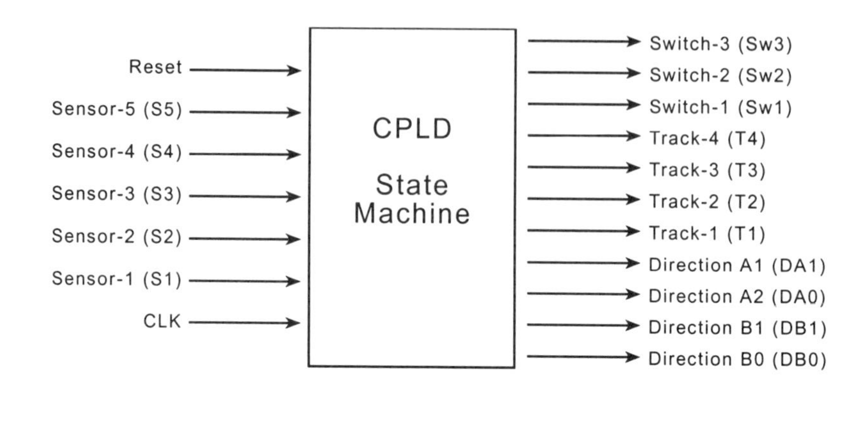

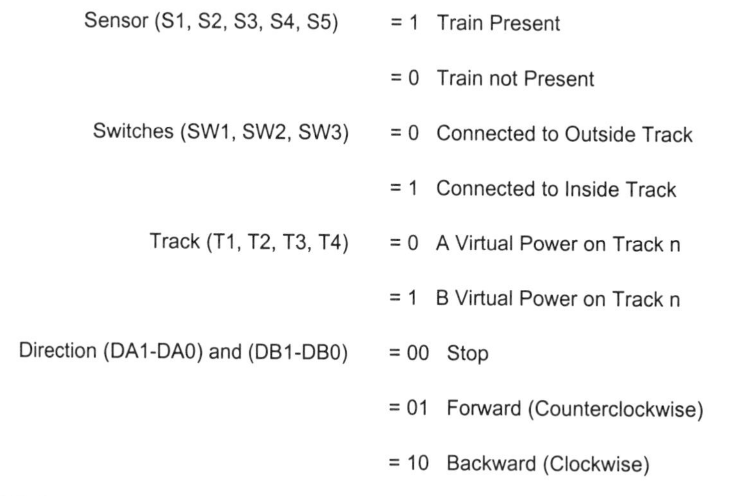

Figure 7.5 State Machine I/O Configuration

7.6 An Example Controller Design

This is a working example of a train controller state machine. For this controller, two trains run counterclockwise at various speeds and avoid collisions. One Train (A) runs on the outer track and the other (B) runs on the inner track. Only one train at a time is allowed to occupy the common track.

Both an ASM chart and a classic state bubble diagram are illustrated in Figures 7.6 and 7.7 respectively. In the ASM chart, state names, ABout, Ain, Bin, Bstop, and Astop indicate the active and possible states. The rectangles contain the active (High) outputs for the given state. Outputs not listed are inactive (Low).

The diamond shapes in the ASM chart indicate where the state machine tests the condition of the inputs (S1, S2, etc.). When two signals are shown in a diamond, they are both tested at the same time for the indicated values.

A state machine classic bubble diagram is shown in Figure 7.7. Both Figures 7.6 and 7.7 contain the same information. They are simply different styles of representing a state diagram. The track diagrams in Figure 7.8 show the states visually. In the state names, "in" and "out" refer to the state of track 2, the track that is common to both loops.

Description of States in Example State Machine

All States

- T3 Asserted: The B power supply is assigned to track 3.
- All signals that are not "Asserted" are zero and imply a logical result as described.

ABout: "Trains A and B Outside"

- DA0 Asserted: Train A is on the outside track and moving counterclockwise (forward).
- DB0 Asserted: Train B is on the inner track (not the common track) and also moving forward.
- Note that by NOT Asserting DA1, it is automatically zero -- same for DB1. Hence, the outputs are DA = "01" and DB = "01".

Ain: "Train A moves to Common Track"

- Sensor 1 has fired either first or at the same time as Sensor 2.
- Either Train A is trying to move towards the common track, or
- Both trains are attempting to move towards the common track.
- Both trains are allowed to enter here; however, state Bstop will stop B if both have entered.
- DA0 Asserted: Train A is on the outside track and moving counterclockwise (forward).
- DB0 Asserted: Train B is on the inner track (not the common track) and also moving forward.

Bstop: "Train B stopped at S2 waiting for Train A to clear common track"

- DA0 Asserted: Train A is moving from the outside track to the common track.
- Train B has arrived at Sensor 2 and is stopped and waits until Sensor 4 fires.
- SW1 and SW2 are NOT Asserted to allow the outside track to connect to common track.
- Note that T2 is not asserted making Track 2 tied to the A Power Supply.

Bin: "Train B has reached Sensor 2 before Train A reaches Sensor 1"

- Train B is allowed to enter the common track. Train A is approaching Sensor 1.
- DA0 Asserted: Train A is on the outside track and moving counterclockwise (forward).
- DB0 Asserted: Train B is on the inner track moving towards the common track.
- SW1 Asserted: Switch 1 is set to let the inner track connect to the common track.
- SW2 Asserted: Switch 2 is set to let the inner track connect to the common track.
- T2 Asserted: The B Power Supply is also assigned to the common track.

Astop: "Train A stopped at S1 waiting for Train B to clear the common track"

- DB0 Asserted: Train B is on the inner track moving towards the common track.
- SW1 and SW2 Asserted: Switches 1 and 2 are set to connect the inner track to the common track.
- T2 Asserted: The B Power Supply is also assigned to the common track.

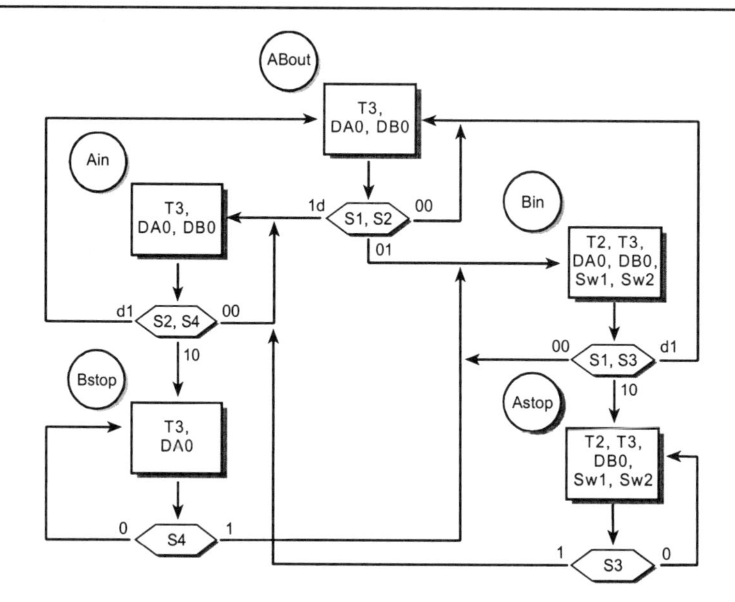

Figure 7.6 Example Train Controller ASM Chart.

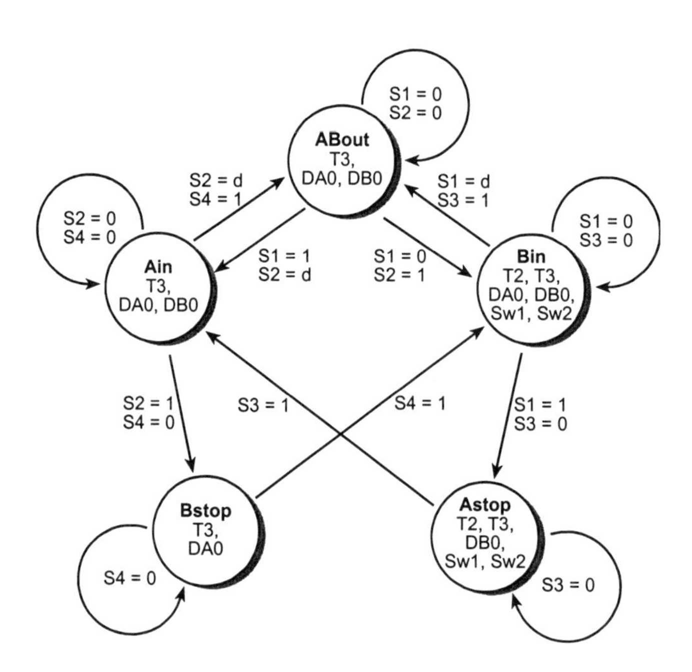

Figure 7.7 Example Train Controller State Diagram.

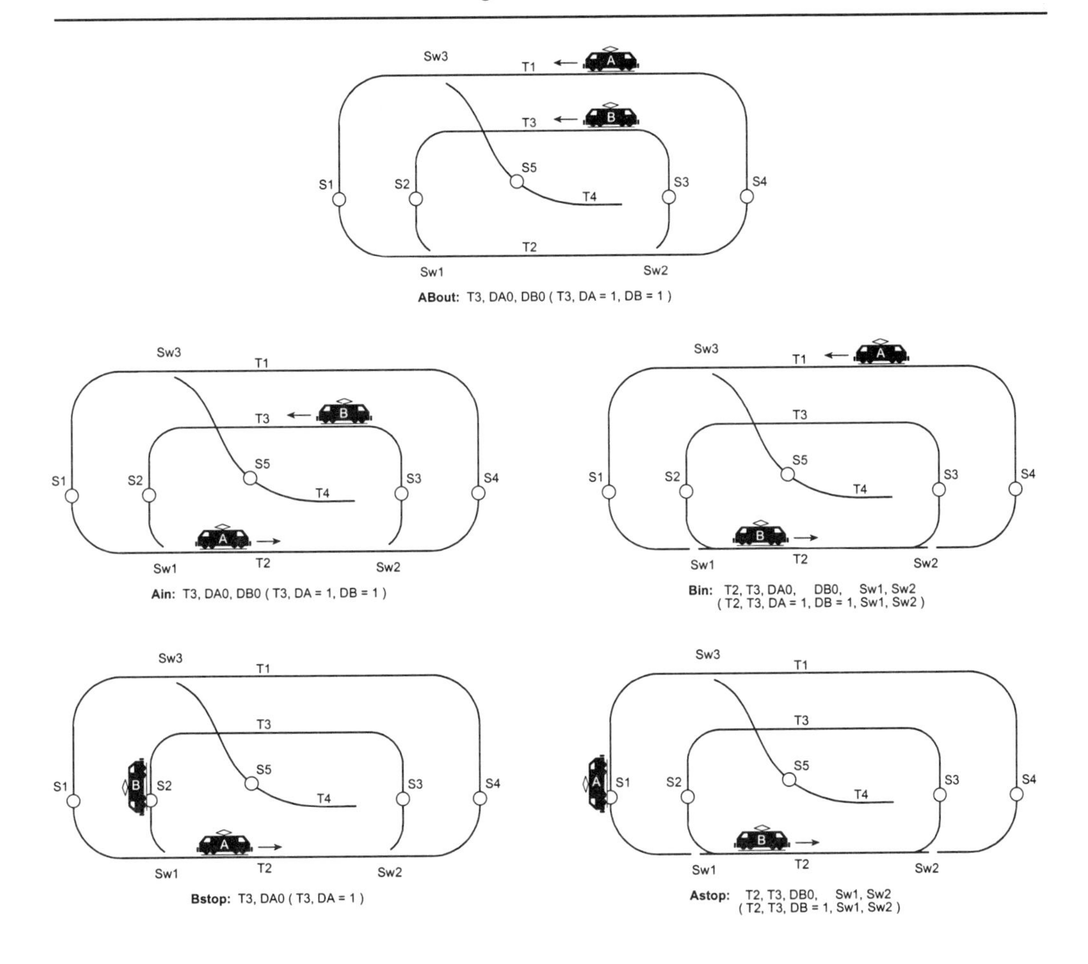

Figure 7.8 Working diagrams of train positions for each state.

Table 7.1 Outputs corresponding to states.

State	ABout	Ain	Astop	Bin	Bstop
Sw1	0	0	1	1	0
Sw2	0	0	1	1	0
Sw3	0	0	0	0	0
T1	0	0	0	0	0
T2	0	0	1	1	0
T3	1	1	1	1	1
T4	0	0	0	0	0
DA(1-0)	01	01	00	01	01
DB(1-0)	01	01	01	01	00

7.7 VHDL Based Example Controller Design

The corresponding VHDL code for the state machine in Figures 7.6 and 7.7 is shown below. A CASE statement based on the current state examines the inputs to select the next state. At each clock edge the next state becomes the current state. WITH...SELECT statements at the end of the program specify the outputs for each state. For additional VHDL help, see the help files in the Altera CAD tools or look at the VHDL examples in Chapter 6.

```
-- Example State machine to control trains-- File: Tcontrol.vhd
--
-- These libraries are required in all VHDL source files
LIBRARY IEEE;
USE IEEE.STD_LOGIC_1164.ALL;
USE IEEE.STD_LOGIC_ARITH.ALL;
USE IEEE.STD_LOGIC_UNSIGNED.ALL;

                    -- This section defines state machine inputs and outputs
                    -- No modifications should be needed in this section
ENTITY Tcontrol IS
PORT( reset, clock, sensor1, sensor2,
        sensor3, sensor4, sensor5       : IN    STD_LOGIC;
        switch1, switch2, switch3       : OUT   STD_LOGIC;
        track1, track2, track3, track4  : OUT   STD_LOGIC;
                    -- dirA and dirB are 2-bit logic vectors(i.e. an array of 2 bits)
        dirA, dirB                      : OUT STD_LOGIC_VECTOR( 1 DOWNTO 0 ));
END Tcontrol;

                        -- This code describes how the state machine operates
                        -- This section will need changes for a different  state machine
ARCHITECTURE a OF Tcontrol IS

                        -- Define local signals (i.e. non input or output signals) here
  TYPE STATE_TYPE IS ( ABout, Ain, Bin, Astop, Bstop );
  SIGNAL state: STATE_TYPE;
  SIGNAL sensor12, sensor13, sensor24   : STD_LOGIC_VECTOR(1 DOWNTO 0);

BEGIN
                        -- This section describes how the state machine behaves
                        -- this process runs once every time reset or the clock changes
  PROCESS ( reset, clock )
  BEGIN
                        -- Reset to this state (i.e. asynchronous reset)
    IF reset = '1' THEN
            state <= ABout;
    ELSIF clock'EVENT AND clock = '1' THEN

                        -- clock'EVENT means value of clock just changed
                        --This section will execute once on each positive clock edge
                        --Signal assignments in this section will generate D flip-flops
                        -- Case statement to determine next state
        CASE state IS
```

```
                    WHEN ABout =>
                                -- This Case checks both sensor1 and sensor2 bits
                        CASE Sensor12 IS
                                -- Note: VHDL's use of double quote for bit vector versus
                                -- a single quote for only one bit!
                                WHEN "00" => state <= About;
                                WHEN "01" => state <= Bin;
                                WHEN "10" => state <= Ain;
                                WHEN "11" => state <= Ain;
                                -- Default case is always required
                                WHEN OTHERS => state <= ABout;
                        END CASE;

                    WHEN Ain =>
                        CASE Sensor24 IS
                                WHEN "00" => state <= Ain;
                                WHEN "01" => state <= ABout;
                                WHEN "10" => state <= Bstop;
                                WHEN "11" => state <= ABout;
                                WHEN OTHERS => state <= ABout;
                        END CASE;

                    WHEN Bin =>
                        CASE Sensor13 IS
                                WHEN "00" => state <= Bin;
                                WHEN "01" => state <= ABout;
                                WHEN "10" => state <= Astop;
                                WHEN "11" => state <= About;
                                WHEN OTHERS => state <= ABout;
                        END CASE;

                    WHEN Astop =>
                        IF Sensor3 = '1' THEN
                                state <= Ain;
                        ELSE
                                state <= Astop;
                        END IF;

                    WHEN Bstop =>
                        IF Sensor4 = '1' THEN
                                state <= Bin;
                        ELSE
                                state <= Bstop;
                        END IF;
                END CASE;
        END IF;
END PROCESS;
                            -- combine sensor bits for case statements above
                            -- "&" operator combines bits
sensor12 <= sensor1 & sensor2;
sensor13 <= sensor1 & sensor3;
sensor24 <= sensor2 & sensor4;
```

```
                                -- These outputs do not depend on the state
Track1  <= '0';
Track4  <= '0';
Switch3 <= '0';
                          -- Outputs that depend on state, use state to select value
                          -- Be sure to specify every output for every state
                          -- values will not default to zero!
WITH state SELECT
        Track3    <=     '1'     WHEN ABout,
                         '1'     WHEN Ain,
                         '1'     WHEN Bin,
                         '1'     WHEN Astop,
                         '1'     WHEN Bstop;
WITH state SELECT
        Track2    <=     '0'     WHEN ABout,
                         '0'     WHEN Ain,
                         '1'     WHEN Bin,
                         '1'     WHEN Astop,
                         '0'     WHEN Bstop;
WITH state SELECT
        Switch1   <=     '0'     WHEN ABout,
                         '0'     WHEN Ain,
                         '1'     WHEN Bin,
                         '1'     WHEN Astop,
                         '0'     WHEN Bstop;
WITH state SELECT
        Switch2   <=     '0'     WHEN ABout,
                         '0'     WHEN Ain,
                         '1'     WHEN Bin,
                         '1'     WHEN Astop,
                         '0'     WHEN Bstop;
WITH state SELECT
        DirA      <=     "01"    WHEN ABout,
                         "01"    WHEN Ain,
                         "01"    WHEN Bin,
                         "00"    WHEN Astop,
                         "01"    WHEN Bstop;
WITH state SELECT
        DirB      <=     "01"    WHEN ABout,
                         "01"    WHEN Ain,
                         "01"    WHEN Bin,
                         "01"    WHEN Astop,
                         "00"    WHEN Bstop;
END a;
```

7.8 Simulation Vector file for State Machine Simulation

The text file, *tcontrol.vec*, controls the simulation and tests the state machine. A vector file is an alternative way to specify simulation stimulus. This file sets up a 40ns clock and specifies sensor patterns (inputs to the state machine), which will be used to test the state machine. The numbers left of the ">" are the time at which the patterns change. As an example "INPUT Sensor1, PATTERN,

0>0" means that the Sensor1 input is a zero at time zero and "100>1" would change it to a '1' at time 100ns in the simulation.

These patterns were chosen by picking a path in the state diagram that moves to all of the different states. The sensor-input patterns would need to be changed to test a different state machine. Sensor inputs should not change faster than the clock cycle time of 40ns. As a minimum, try to test all states and arcs in your state machine simulation.

```
                        % Simulation Test Vectors for Train Control State Machine %
                        %    - File: Tcontrol.vec %

                        % Simulation Start time, Stop time, and Interval%
START 0us;
STOP 1100ns;
INTERVAL 20ns;

                        % Specify the values of all inputs in the simulation %
                        % Setup the System Clock %
INPUTS Clock;
PATTERN
0 1;
                        % Force a Reset to Start %
INPUTS Reset;
PATTERN
                        % Pattern Format is time>value %
0>1
100>0;
                        % Cycle Sensor Inputs at following times %
                        %         to cycle through all states%

                        %Specify Sensor1 values%
INPUTS Sensor1;
PATTERN
0>0
100>0
200>1
300>0
400>0
500>0
600>0
700>0
800>1;
                        %Specify Sensor2 values%
INPUTS Sensor2;
PATTERN
0>0
100>0
200>0
300>1
400>0
500>0
```

```
600>0
700>1;

                    %Specify Sensor3 values%

INPUTS Sensor3;
PATTERN
0>0
100>0
200>0
300>0
400>0
500>0
600>1
700>0
800>0
900>1
1000>0;

                    %Specify Sensor4 values%

INPUTS Sensor4;
PATTERN
0>0
100>0
200>0
300>0
400>1
500>0
600>0
700>0
800>0
900>0
1000>1;

                    % Display these output signals on simulation timing diagram %
                    % Buried means it is an internal signal - not an input or output %
BURIED state;
OUTPUTS switch1 switch2 switch3 track1 track2 track3 track4 dirA dirB;
```

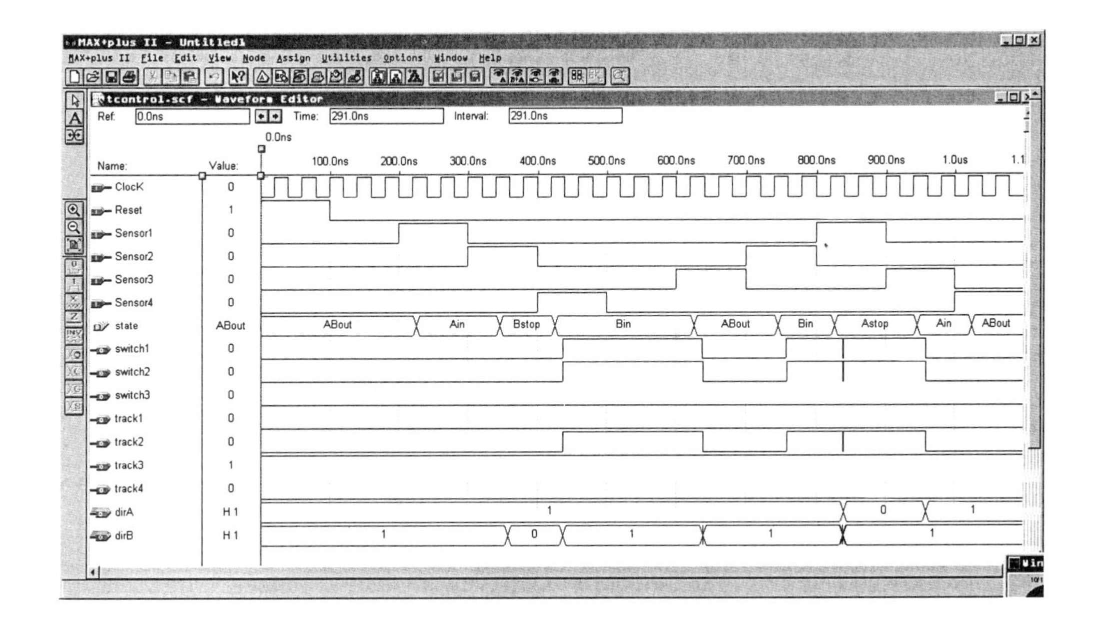

Figure 7.9 Simulation of Tcontrol.vhd using Tcontrol.vec vector file.

7.9 Running the Train Control Simulation

Follow these steps to compile and simulate the state machine for the electric train controller.

Select Current Project

Make Tcontrol.vhd the current project with **File ⇨ Project ⇨ Name**

Then find and select **Tcontrol.vhd**.

Compile and Simulate

Select **File ⇨ Project ⇨ Save Compile and Simulate**.

The simulator will run automatically if there are no compile errors. Select **Open SCF** to see the timing diagram display of your simulation as seen in Figure 7.9. Whenever you change your VHDL source you need to repeat this step. If you get compile errors, clicking on the error will move the text editor to the error location. The Altera software has extensive online help including VHDL syntax examples.

Make any text changes to Tcontrol.vhd or Tcontrol.vec (test vector file) with **File ⇨ Open**. This brings up a special editor window. Note that the menus at the top of the screen change depending on which window is currently open.

Updating new Simulation Test Vectors

To update the simulation with new test vectors from a modified Tcontrol.vec file, select **File ⇨ Open Tcontrol.scf** or select the timing display window if it is already open. If you have changed Tcontrol.vec, update the scf file with **File ⇨ Import Vector File**. You can then hit Start on the simulation window to rerun the simulation with new test vectors.

7.10 Running the Video Train System (After Successful Simulation)

A simulated or "virtual" train system is provided to test the controller without putting trains and people at risk. The simulation runs on the FLEX chip. The output of the simulation is displayed on a VGA monitor connected directly to the UP 1 board. A typical video output display is seen in Figure 7.10. This module is also written in VHDL and it provides the sensor inputs and uses the outputs from the state machine to control the trains. The module tcontrol.vhd is automatically connected to the train simulation.

Here are the steps to run the virtual train system simulation:

Select the top-level project

Make Train.vhd the current project with **File ⇨ Project ⇨ Name**

Then find and select **Train.vhd**. Train.acf must be in the project directory since it contains the FLEX chip pin assignment information.

Compile the Project

Select **File ⇨ Project ⇨ Save and Compile**. Train.vhd will link in your tcontrol.vhd file if it is in the same directory, when compiled. This is a large program so it will take a few minutes to compile. Try running on a fast machine with 32M or more of memory.

Download the Video Train Simulation

Select **MaxPlus ⇨ Programmer**. Pull down the JTAG menu at top to see that Multi-Device is turned on. In **JTAG ⇨ Multi Device JTAG chain setup** select **EPF10K20** or for a new UP-1X board **EPF10K70** as the device and **train.sof** as the programming file to download. In case of problems, see the UP 1 board tutorial in Chapter 1 for more details. Under **Options ⇨ Hardware** select **Byteblaster** and **Lpt1**. Note: Jumpers must be set for FLEX only on UP 1 board, the power supply must be connected, and the Byteblaster* cable must be plugged into the PC's printer port. When everything is setup, the configure button in the programming window should highlight. To download the board, click on the highlighted **configure** button. Attach a VGA monitor to the UP 1 board.

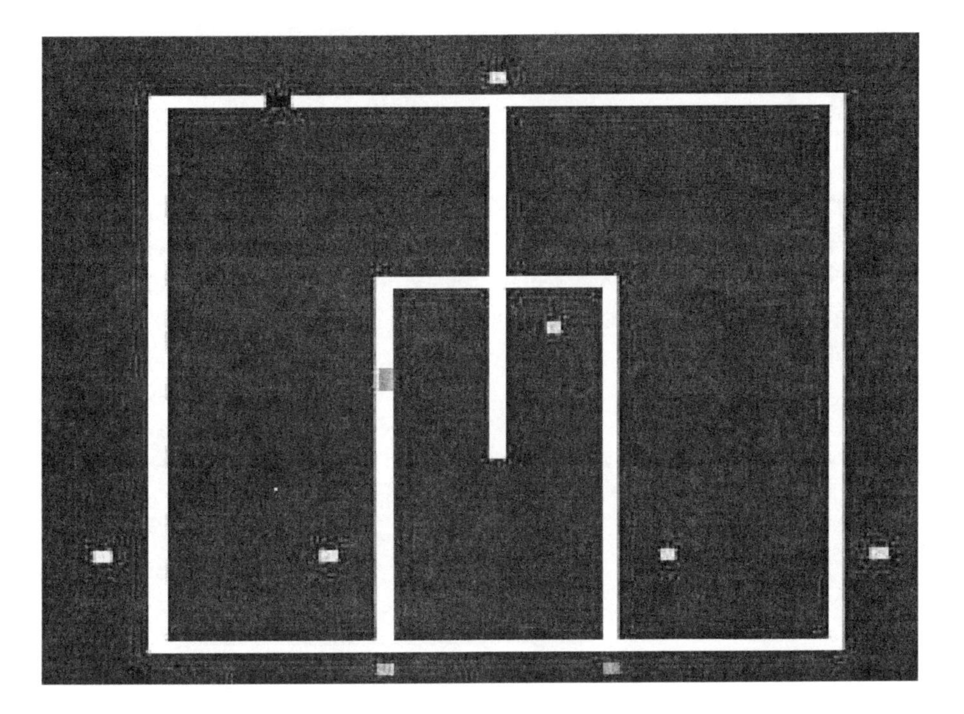

Figure 7.10 Video Image from Train System Simulation.

Viewing the Video Train Simulation

Train output should appear on the VGA monitor after downloading is complete. Flex PB1 is run/step and Flex PB2 is the reset. Train A is displayed in black and Train B is displayed in red.

Sensor and switch values are indicated with a green or red square on the display. Switch values are the squares next to each switch location. Green indicates no train present on a sensor and it indicates switch connected to outside track for a switch. The left FLEX seven-segment display shows the values of the track signals in hexadecimal and the right display indicates the values of DA and DB in hexadecimal.

If a train wreck is detected, the simulation halts and the monitor will flash. The FLEX DIP switches control the speed of Train A (low 4 bits) and B (high 4 bits). Be sure to check operation with different train speeds. Most problems occur with a fast and a slow train.

7.11 Laboratory Exercises

1. Assuming that train A now runs clockwise and B remains counterclockwise, draw a new state diagram and implement the new controller. If you use VHDL to design the new controller, you can modify the code presented in section 7.7. Simulate the controller and then run the video train simulation.

2. Design a state machine to operate the two trains avoiding collisions but minimizing their idle time. Trains must not crash by moving the wrong direction into an open switch. Develop a simulation to verify your state machine is operating correctly before running the video train system.

   The trains are assumed to be in the initial positions as shown in Figure 7.11. Train A is to move counterclockwise around the outside track until it comes to Sensor 1, then move to the inside track stopping at Sensor 5 and waiting for B to pass Sensor 3 twice. Trains can move at different speeds so no assumption should be made about the train speeds. A train hitting a sensor can be stopped before entering the switch area.

   Once B has passed Sensor 3 twice, Train A moves to the outside track and continues around counterclockwise until it picks up where it left off at the starting position as shown in Figure 7.12. Train B is to move as designated only stopping at a sensor to avoid collisions with A.

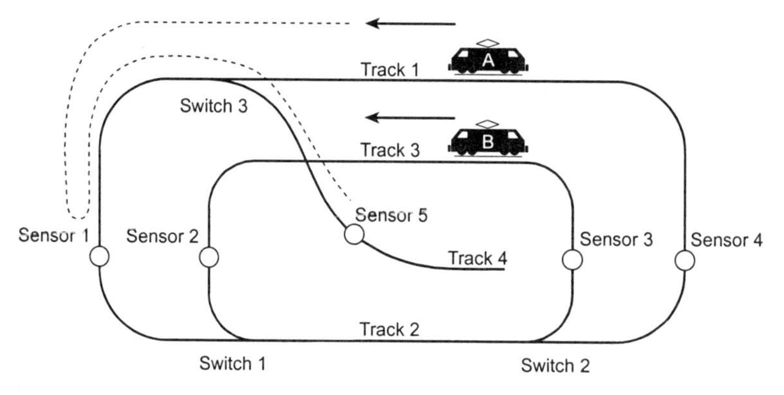

Figure 7.11 Initial Positions of Trains at State Machine Reset with Initial Paths Designated.

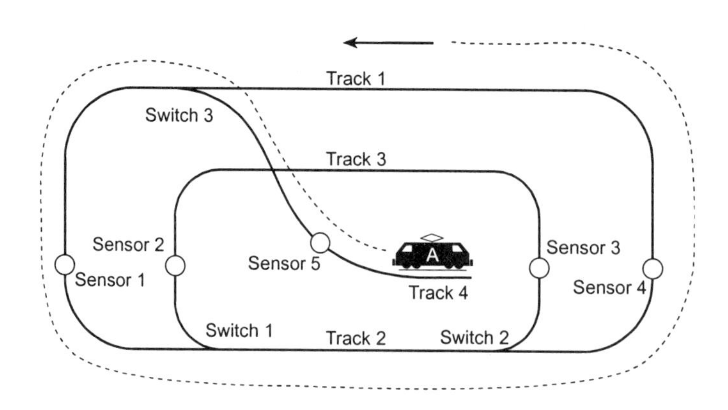

Figure 7.12 Return Path of Train A.

Train B will then continue as soon as there is no potential collision and continue as designated. Trains A and B should run continuously, stopping only to avoid a potential collision.

3. Use the single pulse UP1core functions on each raw sensor input to produce state machine sensor inputs that go High for only one clock cycle per passage of a train. Rework the state machine design with this assumption and repeat problem 1 or 2.
4. Develop another pattern of train movement and design a state machine to implement it.

CHAPTER 8

A Simple Computer Design: The μP 1

8 A Simple Computer Design: The μP 1

A traditional digital computer consists of three main units, the processor or central processing unit (CPU), the memory that stores program instructions and data, and the input/output hardware that communicates to other devices. As seen in Figure 8.1, these units are connected by a collection of parallel digital signals called a bus. Typically, signals on the bus include the memory address, memory data, and bus status. Bus status signals indicate the current bus operation, memory read, memory write, or input/output operation.

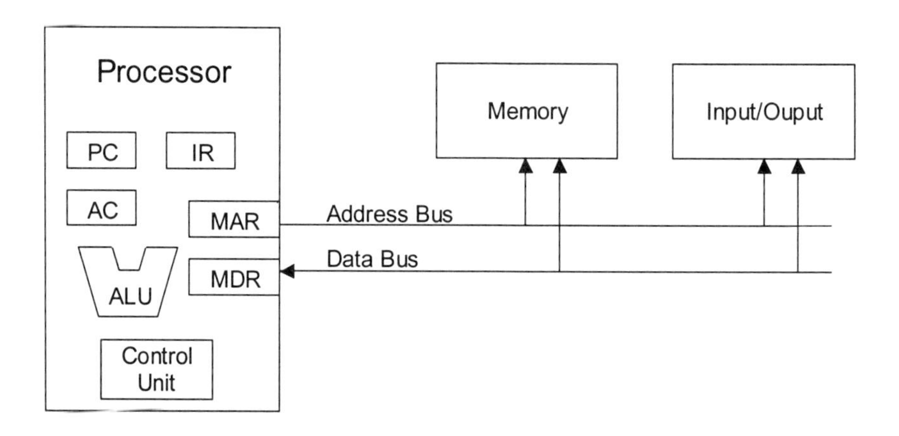

Figure 8.1 Architecture of a Simple Computer System.

Internally, the CPU contains a small number of registers that are used to store data inside the processor. Registers such as PC, IR, AC, MAR and MDR are built using D flip-flops for data storage. One or more arithmetic logic units (ALUs) are also contained inside the CPU. The ALU is used to perform arithmetic and logical operations on data values. Common ALU operations include add, subtract, and logical and/or operations. Register-to-bus connections are hard wired for simple point-to-point connections. When one of several registers can drive the bus, the connections are constructed using multiplexers, open collector outputs, or tri-state outputs. The control unit is a complex state machine that controls the internal operation of the processor.

The primary operation performed by the processor is the execution of sequences of instructions stored in main memory. The CPU or processor reads or fetches an instruction from memory, decodes the instruction to determine what operations are required, and then executes the instruction. The control unit controls this sequence of operations in the processor.

8.1 Computer Programs and Instructions

A computer program is a sequence of instructions that perform a desired operation. Instructions are stored in memory. For the following simple µP 1 computer design, an instruction consists of 16 bits. As seen in Figure 8.2 the high eight bits of the instruction contain the opcode. The instruction operation code or "opcode" specifies the operation, such as add or subtract, that will be performed by the instruction. Typically, an instruction sends one set of data values through the ALU to perform this operation. The low eight bits of each instruction contain a memory address field. Depending on the opcode, this address may point to a data location or the location of another instruction. Some example instructions are shown in Figure 8.3.

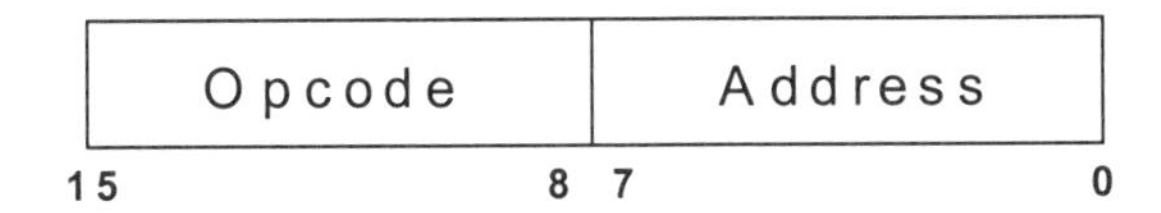

Figure 8.2 Simple µP 1 Computer Instruction Format.

Instruction Mnemonic		**Operation Preformed**	**Opcode Value**
ADD	***address***	**AC <= AC + contents of memory address**	00
STORE	***address***	**contents of memory address <= AC**	01
LOAD	***address***	**AC <= contents of memory address**	02
JUMP	***address***	**PC <= address**	03
JNEG	***address***	**If AC < 0 Then PC <= address**	04

Figure 8.3 Basic µP 1 Computer Instructions.

An example program to compute A = B + C is shown in Figure 8.4. This program is a sequence of three instructions. Program variables such as A, B, and C are typically stored in dedicated memory locations. The symbolic representation of the instructions, called assembly language, is shown in the first column. The second column contains the same program in machine language (the binary pattern that is actually loaded into the computer's memory).

The machine language can be derived using the instruction format in Figure 8.2. First, find the opcode for each instruction in the first column of Figure 8.3. This provides the first two hexadecimal digits in machine language. Second, assign the data values of A, B, and C to be stored in hexadecimal addresses 10,11, and 12 in memory. The address provides the last two hexadecimal digits of each machine instruction.

Assembly Language	Machine Language
LOAD B	**0211**
ADD C	**0012**
STORE A	**0110**

Figure 8.4 Example Computer Program for A = B + C.

The assignment of the data addresses must not conflict with instruction addresses. Normally, the data is stored in memory after all of the instructions in the program. In this case, if we assume the program starts at address 0, the three instructions will use memory addresses 0,1, and 2.

The instructions in this example program all perform data operations and execute in strictly sequential order. Instructions such as JUMP and JNEG are used to transfer control to a different address. Jump and Branch instructions do not execute in sequential order. Jump and Branch instructions must be used to implement control structures such as an IF...THEN statement or program loops. Details are provided in an exercise at the end of this section.

Assemblers are computer programs that automatically convert the symbolic assembly language program into the binary machine language. Compilers are programs that automatically translate higher-level languages, such as C or Pascal, into a sequence of machine instructions. Many compilers also have an option to output assembly language to aid in debugging.

The programmer's view of the computer only includes the registers (such as the program counter) and details that are required to understand the function of assembly or machine language instructions. Other registers and control hardware, such as the instruction register (IR), memory address register (MAR), and memory data register (MDR), are internal to the CPU and are not described in the assembly language level model of the computer. Computer engineers designing the processor must understand the function and operation of these internal registers and additional control hardware.

8.2 The Processor Fetch, Decode and Execute Cycle

The processor reads or fetches an instruction from memory, decodes the instruction to determine what operations are required, and then executes the instruction as seen in Figure 8.5. A simple state machine called the control unit controls this sequence of operations in the processor. The fetch, decode, and execute cycle is found in machines ranging from microprocessor-based PCs to supercomputers. Implementation of the fetch, decode, and execute cycle requires several register transfer operations and clock cycles in this example design.

The program counter contains the address of the current instruction. Normally, to fetch the next instruction from memory the processor must increment the program counter (PC). The processor must then send the address value in the PC to memory over the bus by loading the memory address register (MAR) and start a memory read operation on the bus. After a small delay, the instruction data will appear on the memory data bus lines, and it will be latched into the memory data register (MDR).

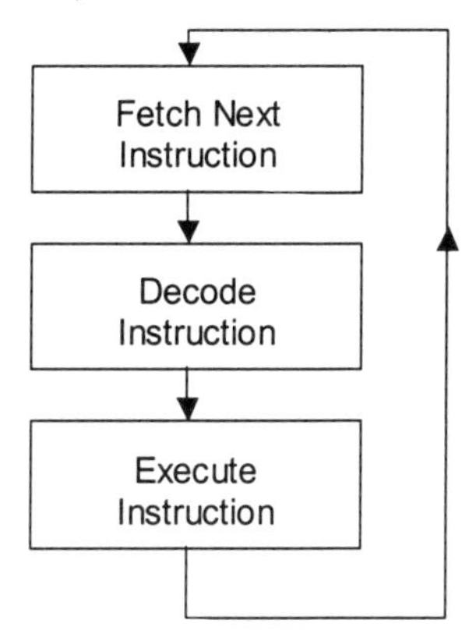

Figure 8.5 Processor Fetch, Decode and Execute Cycle.

Execution of the instruction may require an additional memory cycle so the instruction is normally saved in the CPU's instruction register (IR). Using the value in the IR, the instruction can now be decoded. Execution of the instruction will require additional operations in the CPU and perhaps additional memory operations.

The Accumulator (AC) is the primary register used to perform data calculations and to hold temporary program data in the processor. After completing execution of the instruction the processor begins the cycle again by fetching the next instruction.

The detailed operation of a computer is often modeled by describing the register transfers occurring in the computer system. A variety of register transfer level (RTL) languages such as VHDL or Verilog are designed for this application. Unlike more traditional programming languages, RTL languages can model parallel operations and map easily into hardware designs. Logic synthesis tools can also be used to implement a hardware design automatically using an RTL description.

To explain the function and operation of the CPU in detail, consider the example computer design in Figure 8.1. The CPU contains a general-purpose data register called the accumulator (AC) and the program counter (PC). The arithmetic logic unit (ALU) is used for arithmetic and logical operations.

The fetch, decode, and execute cycle can be implemented in this computer using the sequence of register transfer operations shown in Figure 8.6. The next instruction is fetched from memory with the following register transfer operations:

MAR = PC

Read Memory, MDR = Instruction value from memory

IR = MDR

PC = PC + 1

After this sequence of operations, the current instruction is in the instruction register (IR). This instruction is one of several possible machine instructions such as ADD, LOAD, or STORE. The opcode field is tested to decode the specific machine instruction. The address field of the instruction register contains the address of possible data operands. Using the address field, a memory read is started in the decode state.

The decode state transfers control to one of several possible next states based on the opcode value. Each instruction requires a short sequence of register transfer operations to implement or execute that instruction. These register transfer operations are then performed to execute the instruction. Only a few of the instruction execute states are shown in Figure 8.6. When execution of the current instruction is completed, the cycle repeats by starting a memory read operation and returning to the fetch state. A small state machine called a control unit is used to control these internal processor states and control signals.

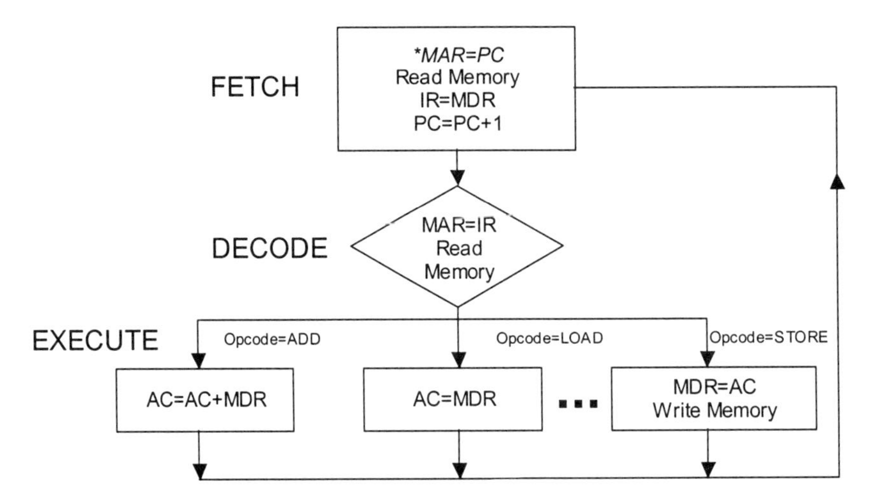

Figure 8.6 Detailed View of Fetch, Decode, and Execute for the μP 1 Computer Design.

Figure 8.7 is the datapath used for the implementation of the μP 1 Computer. A computer's datapath consists of the registers, memory interface, ALUs, and the bus structures used to connect them. The vertical lines are the three major busses used to connect the registers. On the bus lines in the datapath, a "/" with a number indicates the number of bits on the bus. Data values present on the active busses are shown in hexadecimal. MW is the memory write control line.

A reset must be used to force the processor into a known state after power is applied. The initial contents of registers and memory produced by a reset can also be seen in Figure 8.7. Since the PC and MAR are reset to 00, program execution will start at 00.

Note that memory contains the machine code for the example program presented earlier. Recall that the program consists of a LOAD, ADD, and

STORE instruction starting at address 00. Data values for this example program are stored in memory locations, 10, 11, and 12.

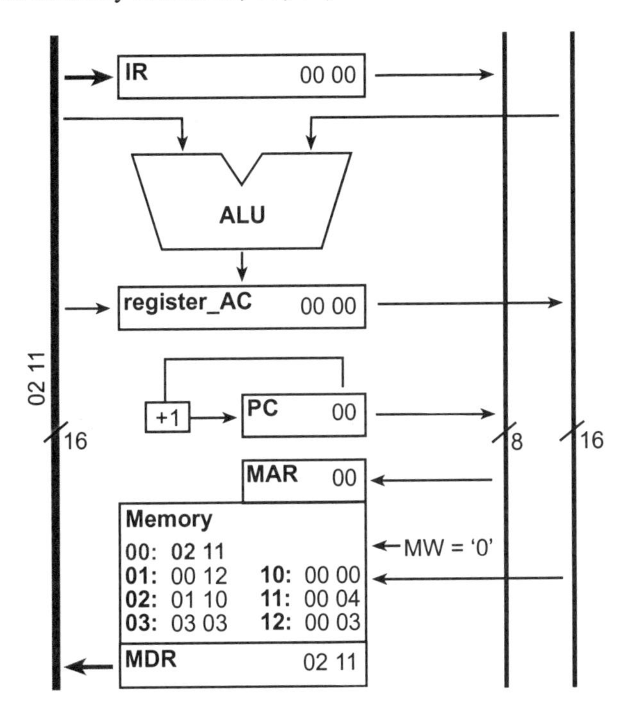

Figure 8.7 Datapath used for the μP 1 Computer Design after applying reset.

Consider the execution of the ADD machine instruction (0012) stored at program location 01 in detail. The instruction, ADD *address*, adds the contents of the memory location at address 12 to the contents of AC and stores the result in AC. The following sequence of register transfer operations will be required to fetch and execute this instruction.

FETCH: *REGISTER TRANSFER CYCLE 1*:

***MAR = PC prior to fetch*, read memory, IR = MDR, PC = PC + 1**

First, the memory address register is loaded with the PC. In the example program, the ADD instruction (0012) is at location 01 in memory, so the PC and MAR will both contain 01. In this implementation of the computer, the MAR=PC operation will be moved to the end of the fetch, decode, and execute loop to the execute state in order to save a clock cycle. To fetch the instruction, a memory read operation is started. After a small delay for the memory access time, the ADD instruction is available at the input of the instruction register. To set up for the next instruction fetch, one is added to the program counter. The last two operations occur in parallel during one clock cycle using two different data busses. At the rising edge of the clock signal, the decode state is entered.

A block diagram of the register transfer operations for the fetch state is seen in Figure 8.8. Inactive busses are not shown.

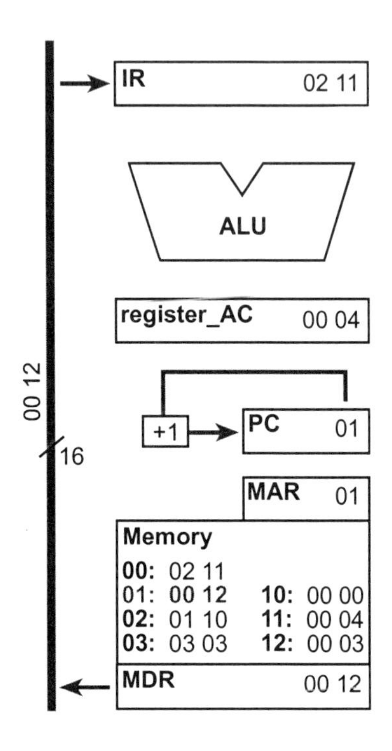

Figure 8.8 Register transfers in the ADD instruction's Fetch State.

DECODE: *REGISTER TRANSFER CYCLE 2*:

Decode Opcode to find Next State, MAR = IR, and start memory read

Using the new value in the IR, the CPU control hardware decodes the instruction's opcode of 00 and determines that this is an ADD instruction. Therefore, the next state in the following clock cycle will be the execute state for the ADD instruction.

Instructions typically are decoded in hardware using combinational circuits such as decoders, programmable logic arrays (PLAs), or perhaps even a small ROM. A memory read cycle is always started in decode, since the instruction may require a memory data operand in the execute state.

The ADD instruction requires a data operand from memory address 12. In Figure 8.9, the low 8–bit address field portion of the instruction in the IR is transferred to the MAR. At the next clock, after a small delay for the memory access time, the ADD instruction's data operand value from memory (0003) will be available in the MDR.

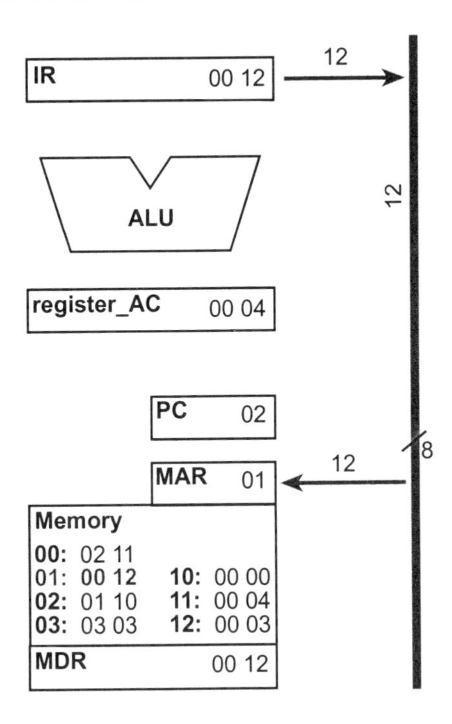

Figure 8.9 Register transfers in the ADD instruction's Decode State.

EXECUTE ADD: ***REGISTER TRANSFER CYCLE 3*****:**
AC = AC + MDR, MAR = PC*, and GOTO FETCH

The two values can now be added. The ALU operation input is set for addition by the control unit. As shown in Figure 8.10, the MDR's value of 0003 is fed into one input of the ALU. The contents of register AC (0004) are fed into the other ALU input. After a small delay for the addition circuitry, the sum of 0007 is produced by the ALU and will be loaded into the AC at the next clock. To provide the address for the next instruction fetch, the MAR is loaded with the current value of the PC (02). Note that by moving the operation, MAR=PC, to every instruction's final execute state, the fetch state can execute in one clock cycle. The ADD instruction is now complete and the processor starts to fetch the next instruction at the next clock cycle. Since three states were required, an ADD instruction will require three clock cycles to complete the operation.

After considering this example, it should be obvious that a thorough understanding of each instruction, the hardware organization, busses, control signals, and timing is required to design a processor. Some operations can be performed in parallel, while others must be performed sequentially. A bus can only transfer one value per clock cycle and an ALU can only compute one value per clock cycle, so ALUs, bus structures, and data transfers will limit those operations that can be done in parallel during a single clock cycle. In the states examined, a maximum of three buses were used for register transfers.

Timing in critical paths, such as ALU delays and memory access times, will determine the clock speed at which these operations can be performed.

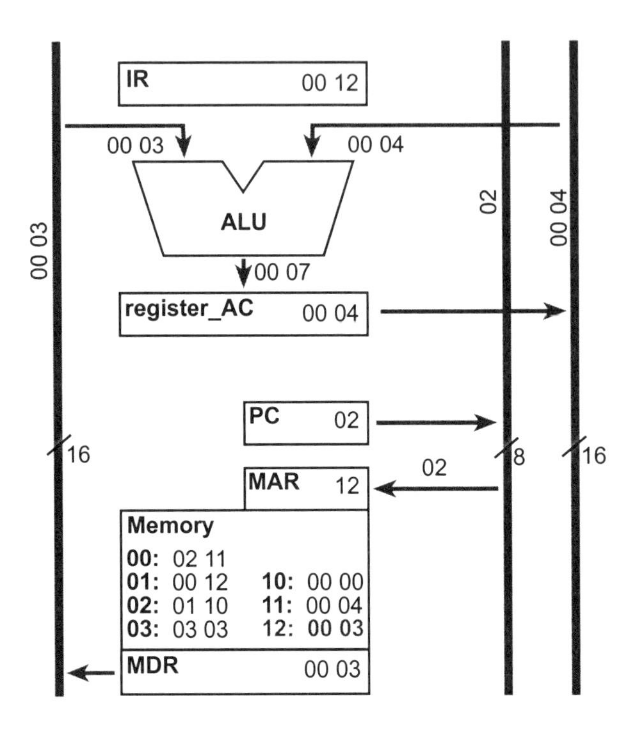

Figure 8.10 Register transfers in the ADD instruction's Execute State.

The μP 1's multiple clock cycles per instruction implementation approach was used in early generation microprocessors. These computers had limited hardware, since the VLSI technology at that time supported orders of magnitude fewer gates on a chip than is now possible in current devices. Current generation processors, such as those used in personal computers, have a hundred or more instructions, and use additional means to speedup program execution. Instruction formats are more complex with up to 32 data registers and with additional instruction bits that are used for longer address fields and more powerful addressing modes.

Pipelining converts fetch, decode, and execute into a parallel operation mode instead of sequential. As an example, with three stage pipelining, the fetch unit fetches instruction n + 2, while the decode unit decodes instruction n + 1, and the execute unit executes instruction n. With this faster pipelined approach, an instruction finishes execution every clock cycle rather than three as in the simple computer design presented here.

Superscalar machines are pipelined computers that contain multiple fetch, decode and execute units. Superscalar computers can execute several instructions in one clock cycle. Most current generation processors including those in personal computers are both pipelined and superscalar. An example of a pipelined, reduced instruction set computer (RISC) design can be found in Chapter 13.

8.3 VHDL Model of the μP 1

To demonstrate the operation of a computer, a VHDL model of the μP 1 computer is shown in Figure 8.11. The simple μP 1 computer design fits easily into a FLEX 10K20 device. The computer's RAM memory is implemented using the LPM_RAM_DQ function.

The machine language program shown in Figure 8.12 is loaded into memory using a memory initialization file (*.mif). This produces 256 words of 16-bit memory for instructions and data. The memory initialization file, program.mif can be edited to change the loaded program. A write is performed only when the memory_write signal is High. On a FLEX 10K20 device, the access time for memory operations is in the range of 20-50ns.

The remainder of the computer model is basically a VHDL-based state machine that implements the fetch, decode, and execute cycle. The first few lines declare internal registers for the processor along with the states needed for the fetch, decode and execute cycle. A long CASE statement is used to implement the control unit state machine. A reset state is needed to initialize the processor. In the reset state, several of the registers are reset to zero and a memory read of the first instruction is started. This forces the processor to start executing instructions at location 00 in a predictable state after a reset.

The fetch state adds one to the PC and loads the instruction into the instruction register (IR). After the rising edge of the clock signal, the decode state starts. In decode, the low eight bits of the instruction register are used to start a memory read operation in case the instruction needs a data operand from memory. The decode state contains another CASE statement to decode the instruction using the opcode value in the high eight bits of the instruction. This means that the computer can have up to 256 different instructions, although only four are implemented in the basic model. Other instructions can be added as exercises. After the rising edge of the clock signal, control transfers to an execute state that is specific for each instruction.

Some instructions can execute in one clock cycle and some instructions may take more than one clock cycle. Instructions that write to memory will require more than one state for execute because of memory timing constraints. As seen in the STORE instruction, the memory address and data needs to be stable before and after the memory write signal is High, hence, additional states are used to avoid violating memory setup and hold times. When each instruction finishes the execute state, MAR is loaded with the PC to start the fetch of the next instruction. After the final execute state for each instruction, control returns to the fetch state.

```
LIBRARY IEEE;
USE  IEEE.STD_LOGIC_1164.ALL;
USE  IEEE.STD_LOGIC_ARITH.ALL;
USE  IEEE.STD_LOGIC_UNSIGNED.ALL;
LIBRARY lpm;
USE lpm.lpm_components.ALL;
ENTITY SCOMP IS
PORT(  clock, reset                  : IN STD_LOGIC;
       program_counter_out           : OUT STD_LOGIC_VECTOR( 7 DOWNTO 0 );
       register_AC_out               : OUT STD_LOGIC_VECTOR(15 DOWNTO 0 );
       memory_data_register_out      : OUT STD_LOGIC_VECTOR(15 DOWNTO 0 ));
END SCOMP;
ARCHITECTURE a OF scomp IS
TYPE STATE_TYPE IS ( reset_pc, fetch, decode, execute_add, execute_load, execute_store,
                     execute_store3, execute_store2, execute_jump );
SIGNAL state: STATE_TYPE;
SIGNAL instruction_register, memory_data_register   : STD_LOGIC_VECTOR(15 DOWNTO 0 );
SIGNAL register_AC                                  : STD_LOGIC_VECTOR(15 DOWNTO 0 );
SIGNAL program_counter                              :  STD_LOGIC_VECTOR( 7 DOWNTO 0 );
SIGNAL memory_address_register                      : STD_LOGIC_VECTOR( 7 DOWNTO 0 );
SIGNAL memory_write                                 : STD_LOGIC;
BEGIN
                                 -- Use LPM function for computer's memory (256 16-bit words)
 memory: lpm_ram_dq
   GENERIC MAP (
         lpm_widthad         => 8,
         lpm_outdata         => "UNREGISTERED",
         lpm_indata          => "REGISTERED",
         lpm_address_control => "UNREGISTERED",
                                 -- Reads in mif file for initial program and data values
         lpm_file            => "program.mif",
         lpm_width           => 16 )
   PORT MAP (  data => Register_AC, address => memory_address_register,
               we => memory_write, inclock => clock, q => memory_data_register );

         program_counter_out        <= program_counter;
         register_AC_out            <= register_AC;
         memory_data_register_out   <= memory_data_register;
PROCESS ( CLOCK, RESET )
   BEGIN
   IF reset = '1' THEN
       state <= reset_pc;
   ELSIF clock'EVENT AND clock = '1' THEN
       CASE state IS
                               -- reset the computer, need to clear some registers
       WHEN reset_pc =>
          program_counter            <= "00000000";
          memory_address_register    <= "00000000";
          register_AC                <= "0000000000000000";
          memory_write               <= '0';
          state                      <= fetch;
                               -- Fetch instruction from memory and add 1 to PC
       WHEN fetch =>
          instruction_register       <= memory_data_register;
          program_counter            <= program_counter + 1;
```

```
        memory_write            <= '0';
        state                   <= decode;
                        -- Decode instruction and send out address of any data operands
    WHEN decode =>
        memory_address_register <= instruction_register( 7 DOWNTO 0);
        CASE instruction_register( 15 DOWNTO 8 ) IS
                WHEN "00000000" =>
                  state <= execute_add;
                WHEN "00000001" =>
                  state <= execute_store;
                WHEN "00000010" =>
                  state <= execute_load;
                WHEN "00000011" =>
                  state <= execute_jump;
                WHEN OTHERS =>
                  state <= fetch;
        END CASE;
                        -- Execute the ADD instruction
    WHEN execute_add =>
        register_ac             <= register_ac + memory_data_register;
        memory_address_register <= program_counter;
        state                   <= fetch;
                        -- Execute the STORE instruction
                        -- (needs three clock cycles for memory write)
    WHEN execute_store =>
                        -- write register_A to memory
        memory_write            <= '1';
        state                   <= execute_store2;
                        -- This state ensures that the memory address is
                        -- valid until after memory_write goes inactive
    WHEN execute_store2 =>
        memory_write            <= '0';
        state                   <= execute_store3;
    WHEN execute_store3 =>
        memory_address_register <= program_counter;
        state                   <= fetch;
                        -- Execute the LOAD instruction
    WHEN execute_load =>
        register_ac             <= memory_data_register;
        memory_address_register <= program_counter;
        state <= fetch;
                        -- Execute the JUMP instruction
    WHEN execute_jump =>
        memory_address_register <= instruction_register( 7 DOWNTO 0 );
        program_counter         <= instruction_register( 7 DOWNTO 0 );
        state                   <= fetch;
    WHEN OTHERS =>
        memory_address_register <= program_counter;
        state <= fetch;
    END CASE;
  END IF;
  END PROCESS;
END a;
```

Figure 8.11 VHDL Model of μP 1 Computer.

```
DEPTH = 256;    % Memory depth and width are required    %
WIDTH = 16;     % Enter a decimal number    %

ADDRESS_RADIX = HEX;    % Address and value radixes are optional    %
DATA_RADIX = HEX;       % Enter BIN, DEC, HEX, or OCT; unless    %
                        % otherwise specified, radixes = HEX    %
-- Specify initial data values for memory, format is address : data
CONTENT
    BEGIN
[00..FF]    :    0000;  % Range--Every address from 00 to FF = 0000 %
-- Computer Program for A = B + C
    00      :    0210;  % LOAD A with MEM(10) %
    01      :    0011;  % ADD MEM(11) to A %
    02      :    0112;  % STORE A in MEM(12) %
    03      :    0212;  % LOAD A with MEM(12) check for new value of FFFF %
    04      :    0304;  % JUMP to 04 (loop forever) %
    10      :    AAAA;  % Data Value of B%
    11      :    5555;  % Data Value of C%
    12      :    0000;  % Data Value of A - should be FFFF after running program %
```

Figure 8.12 MIF file containg µP 1 Computer Program.

8.4 Simulation of the µP1 Computer

A simulation output from the VHDL model is seen in Figure 8.13. After a reset, the test program seen in Figure 8.12, loads, adds, and stores a data value to compute A = B + C. The final value is then loaded again to demonstrate that the memory contains the correct value for A. The program then ends with a jump instruction that jumps back to its own address producing an infinite loop.

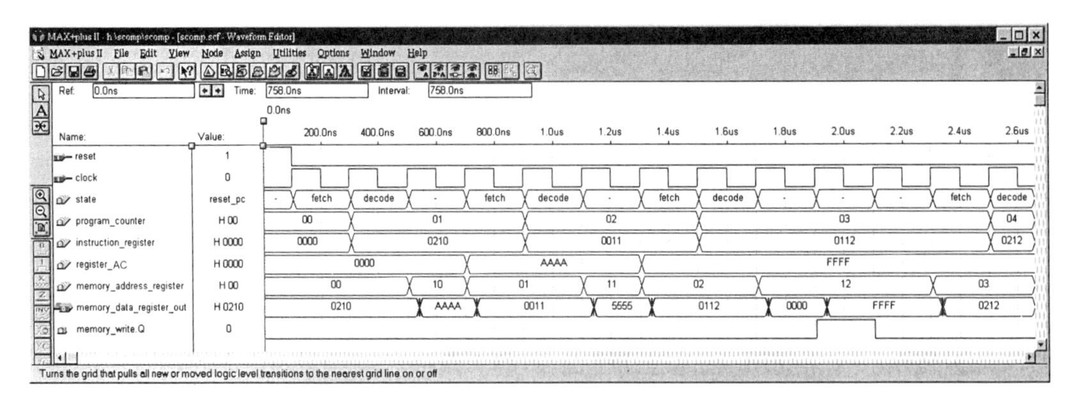

Figure 8.13 Simulation of the Simple µP 1 Computer Program.

8.5 Laboratory Exercises

In these exercises, the meta assembler described in Appendix D can be used. It can save time when writing longer programs for the μP 1 computer.

1. Compile and simulate the μP 1 computer VHDL model. Rewrite the machine language program in the program.mif file to compute A = (B + C) + D. Store D in location 13 in memory. End the program with a Jump instruction that jumps to itself. Be sure to select the FLEX 10K20 (10K70 for a UP 1X board) device as the target, because the computer model is too large for the MAX device. Find the maximum clock rate of the μP 1 computer. Examine the project *.rpt file and find the logic cell (LC) percentage utilized. Multiply by 20,000 (70,000 for a UP 1X board) to estimate the number of gates in the μP 1 computer.

2. Add the JNEG execute state to the CASE statement in the model. JNEG is Jump if AC < 0. If A >= 0 the next sequential instruction is executed. In most cases, a new instruction will just require a new execute state in the decode CASE statement. Use the opcode value of 04 for JNEG. Test the new instruction with the following test program that implements the operation, IF A>= 0 THEN B = C

	Assembly Language		Machine Language	Memory Address
	LOAD	A	0210	00
	JNEG	End_of_If	0404	01
	LOAD	C	0212	02
	STORE	B	0111	03
End_of_If:	JMP	End_of_If	0304	04

End_of_If is an example of a label; it is a symbolic representation for a location in the program. Labels are used in assembly language to mark locations in a program. The last line that starts out with End_of_If: is the address used for the End_of_If symbol in the Jump instruction address field. Assuming the program starts at address 00, the value of the End_of_If label will be 04. Test the JNEG instruction for both cases A < 0 and A >= 0. Place nonzero values in the *.mif file for B and C so that you can verify the program executes correctly.

3. Add the instructions in the table below to the VHDL model, construct a test program for each instruction, compile and simulate to verify correct operation. In JPOS and JZERO instructions, both cases must be tested.

Instruction	Function	Opcode
SUBT *address*	AC = AC - MDR	05
XOR *address*	AC = AC XOR MDR	06
OR *address*	AC = AC OR MDR	07
AND *address*	AC = AC AND MDR	08
JPOS *address*	IF AC > 0 THEN PC = address	09

JZERO *address*	IF AC = 0 THEN PC = address	0A
ADDI *address*	AC = AC + *address*	0B

In the logical XOR instruction each bit is exclusive OR'd with the corresponding bit in each operation for a total of sixteen independent exclusive OR operations. This is called a bitwise logical operation. OR and AND are also bitwise logical operations. The add-immediate instruction, ADDI, sign extends the 8-bit address field value to 16 bits. To sign extend, copy the sign bit to all eight high bits. This allows the use of both positive and negative two's complement numbers for the 8-bit immediate value stored in the instruction.

4. Add the following two shift instructions to the simple computer model and verify with a test program and simulation.

Instruction	Function	Opcode
SHL address	AC = AC shifted left *address* bits	0C
SHR address	AC = AC shifted right *address* bits	0D

The function LPM_CLSHIFT is useful to implement multiple bit shifts. SHL and SHR can also be used if 1993 VHDL features are enabled in the compiler. Only the low four bits of the address field contain the shift amount. The other four bits are always zero.

5. Run the μP 1 computer model using the FLEX 10K20 chip on the UP 1 board or the 10K70 on the UP 1X. Use a debounced pushbutton for the clock and the other pushbutton for reset. With a seven-segment decoder, output the PC in hex to the FLEX seven-segment LED displays. Run a test program on the board and verify the correct value of the PC appears in the seven-segment LED displays by stepping through the program using the pushbutton.

6. Add these two input/output (I/O) instructions to the μP 1 computer model running on the UP 1 board.

Instruction	Function	Opcode
IN i/o address	AC = FLEX DIP switch bits (low 8 bits)	0E
OUT i/o address	Seven-segment displays hex value of AC	0F

These instructions modify or use only the low eight bits of AC. Remove the PC display feature from the previous problem, if it was added. Test the new I/O instructions by writing a program that reads in the switches, adds one to the switch value, and outputs this value to the LED display. Repeat the input, add, and output operation in an infinite loop by jumping back to the start of the program. Add a new register, register_output, to the input of the seven-segment decoder that drives the LED display. The register is loaded with the value of AC only when an OUT instruction is executed. Compile, download, and execute the program on the UP 1 board. When several I/O devices are present, they should respond only to their own unique i/o address, just like memory.

7. Use the timing analyzer to determine the maximum clock rate for the μP 1 computer. Using this value, compute the execution time for the example program in Figure 8.4.

8. Modify the FLEX video output display described in Chapter 9 for the MIPS computer example to display the μP 1's internal registers. While running on the UP 1 board, use the pushbuttons for clock and reset as suggested in problem 5.

9. Add video character output and keyboard input to the computer, after studying the material presented in Chapters 9 and 10.

10. Add the WAIT instruction to the simple computer model and verify with a test program and simulation. WAIT *value*, loads and starts an 8-bit ten-millisecond (10^{-2} second) timer and then waits *value*10* ms before returning to fetch for the next instruction. Use an opcode of 10 for the WAIT instruction.

11. Expand the memory address space of the μP 1 computer from eight bits to nine bits. Some registers will also need an additional bit. Use 512 locations of 16-bit memory. Expand the address field by 1-bit by reducing the size of the opcode field by 1-bit. This will limit the number of instructions to 128 but the maximum program size can now increase from 256 locations to 512 locations.

12. Modify the μP 1 computer so that it uses two different memories. Use one memory for instructions and a new memory for data values. The new data memory should be 256 locations of 16-bit data.

13. Add a subroutine CALL and RETURN instruction to the μP 1 computer design. Use a dedicated register to store the return address or use a stack with a stack pointer register. The stack should start at high addresses and as it grows move to lower addresses.

14. Implement a stack as suggested in the previous problem and add instructions to PUSH or POP register AC from the stack. At reset, set the stack pointer to the highest address of data memory.

15. Add all of the instructions and features suggested in the exercises to the μP 1 computer and use it as a microcontroller core for one of the robot projects suggested in Chapter 12. Additional instructions of your own design along with an interval timer that can be read using the IN instruction may also be useful.

16. Using the two low-bits from the opcode field, add a register address field that selects one of four different data registers A, B, C, or D for each instruction.

17. Use the implementation approach in the μP 1 computer model as a starting point to implement the basic instruction set of a different computer from your digital logic textbook or other reference manual.

CHAPTER 9

VGA Video Signal Generation

MIPS	COMPUTER
PC	00000008
INST	00430820
REG1	00000055
REG2	000000AA
ALU	000000FF
W.B.	000000FF
BRAN	0
ZERO	0
MEMR	0
MEMW	0
CLK	↓
RST	↓

The video image above was produced by a UP 1 board design.

9 VGA Video Display Generation

To understand how it is possible to generate a video image using the Altera UP 1 board, it is first necessary to understand the various components of a video signal. A VGA video signal contains 5 active signals. Two signals compatible with TTL logic levels, horizontal sync and vertical sync, are used for synchronization of the video. Three analog signals with 0.7 to 1.0-Volt peak-to-peak levels are used to control the color. The color signals are Red, Green, and Blue. They are often collectively referred to as the RGB signals. By changing the analog levels of the three RGB signals all other colors are produced.

9.1 Video Display Technology

The technology used to display a video image dictates the very nature of the video signals. The major component inside a VGA computer monitor is the color CRT or Cathode Ray Tube shown in Figure 9.1. The electron beam must be scanned over the viewing screen in a sequence of horizontal lines to generate an image. The deflection yoke uses magnetic or electrostatic fields to deflect the electron beam to the appropriate position on the face of the CRT. The RGB color information in the video signal is used to control the strength of the electron beam. Light is generated when the beam is turned on by a video signal and it strikes a color phosphor dot or line on the face of the CRT. The face of a color CRT contains three different phosphors. One type of phosphor is used for each of the primary colors of red, green, and blue.

In standard VGA format, as seen in Figure 9.2, the screen contains 640 by 480 picture elements or pixels. The video signal must redraw the entire screen 60 times per second to provide for motion in the image and to reduce flicker. This period is called the refresh rate. The human eye can detect flicker at refresh rates less than 30 Hz.

To reduce flicker from interference from fluorescent lighting sources, refresh rates higher than 60 Hz are sometimes used in PC monitors. The onboard clock on the Altera UP 1 board produces a fixed 60Hz refresh rate. The color of each pixel is determined by the value of the RGB signals when the signal scans across each pixel. In 640 by 480-pixel mode, with a 60Hz refresh rate, this is approximately 40 ns per pixel. A 25Mhz clock has a period of 40 ns.

9.2 Video Refresh

The screen refresh process seen in Figure 9.2 begins in the top left corner and paints 1 pixel at a time from left to right. At the end of the first row, the row increments and the column address is reset to the first column. Each row is painted until all pixels have been displayed. Once the entire screen has been painted, the refresh process begins again.

The video signal paints or refreshes the image using the following process. The vertical sync signal, as shown in Figure 9.3 tells the monitor to start displaying a new image or frame, and the monitor starts in the upper left corner with pixel

0,0. The horizontal sync signal, as shown in Figure 9.4, tells the monitor to refresh another row of 640 pixels.

After 480 rows of pixels are refreshed with 480 horizontal sync signals, a vertical sync signal resets the monitor to the upper left comer and the process continues. During the time when pixel data is not being displayed and the beam is returning to the left column to start another horizontal scan, the RGB signals should all be set to the color black (all zeros).

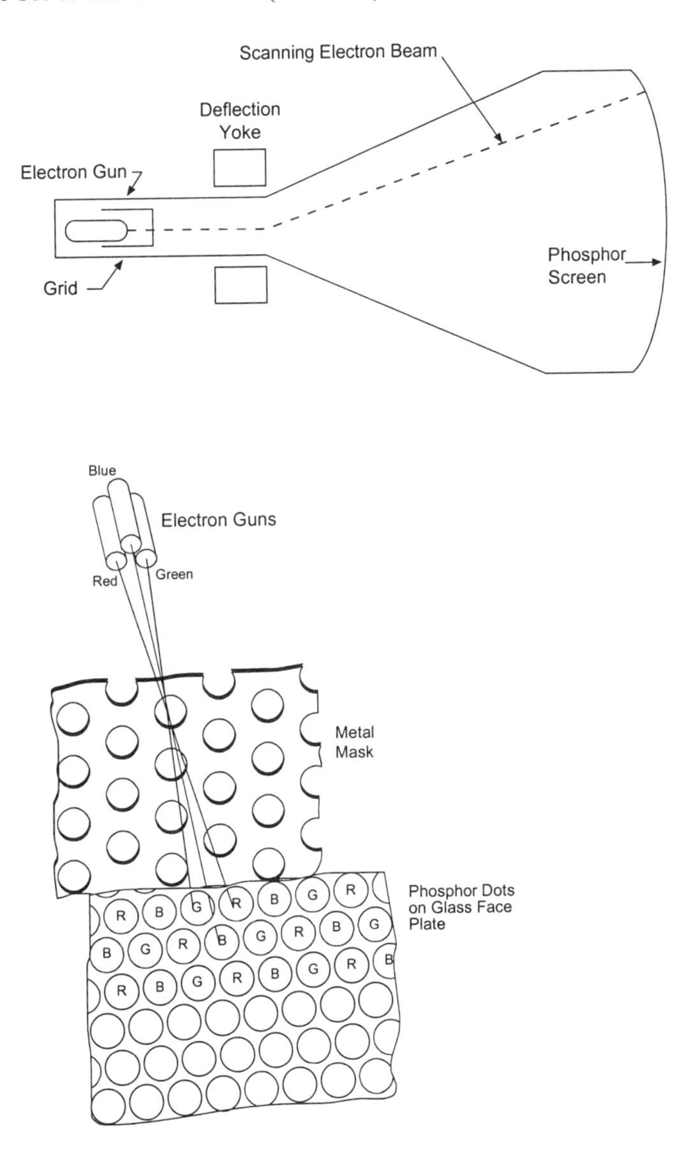

Figure 9.1 Color CRT and Phosphor Dots on Face of Display.

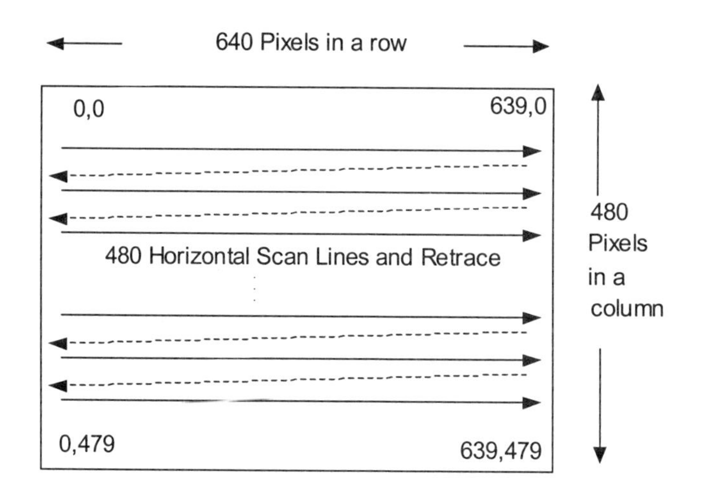

Figure 9.2 VGA Image - 640 by 480 Pixel Layout.

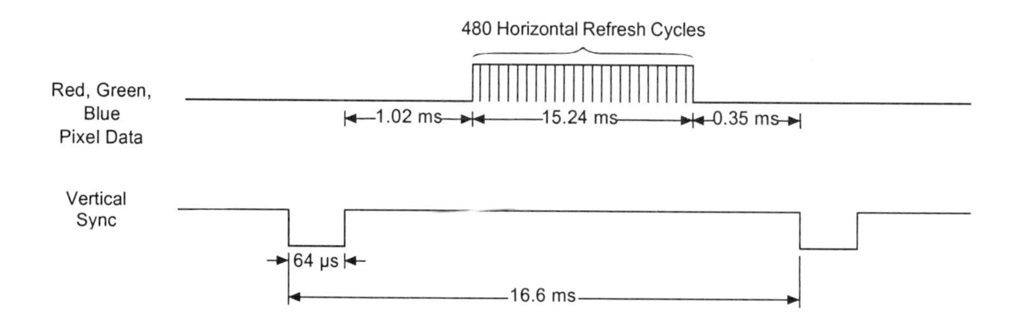

Figure 9.3 Vertical Sync Signal Timing.

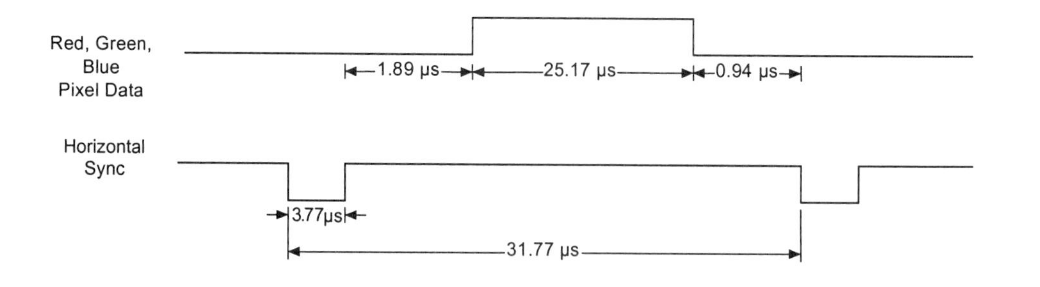

Figure 9.4 Horizontal Sync Signal Timing.

Many VGA monitors will shut down if the two sync signals are not the correct values. Most PC monitors have an LED that is green when it detects valid sync signals and yellow when it does not lock in with the sync signals. In a PC graphics card, a dedicated video memory location is used to store the color value of every pixel in the display. This memory is read out as the beam scans across the screen to produce the RGB signals. There is not enough memory inside current generation FPLD chips for this approach so other techniques will be developed which require less memory.

9.3 Using a CPLD for VGA Video Signal Generation

To provide interesting output options in complex designs, video output can be developed using hardware inside the CPLD or FPGA. Only five signals or pins are required, two sync signals and three RGB color signals. A simple resistor and diode circuit is used to convert TTL output pins from the CPLD to the analog RGB signals for the video signal. This supports two levels for each signal in the RGB data and thus produces a total of eight colors. This circuit and a VGA connector for a monitor are already installed on the Altera UP 1 board.

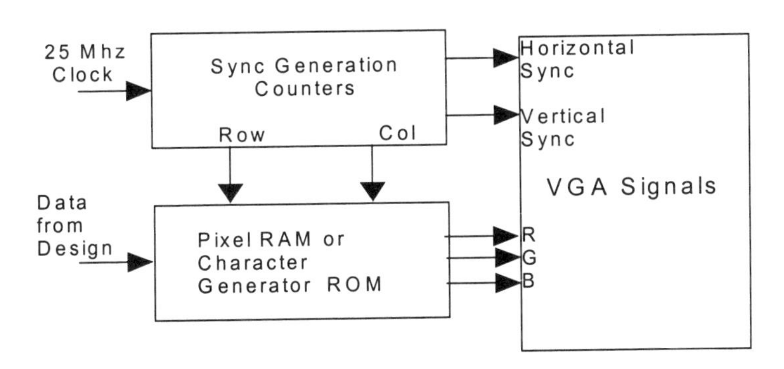

Figure 9.5 CLPD based generation of VGA Video Signals.

As seen in Figure 9.5, a 25.175 MHz clock, which is the 640 by 480 VGA pixel data rate of approximately 40ns is used to drive counters that generate the horizontal and vertical sync signals. Additional counters generate row and column addresses. In some designs, pixel resolution will be reduced from 640 by 480 to a lower resolution by using a clock divide operation on the row and column counters. The row and column addresses feed into a pixel RAM for graphics data or a character generator ROM when used to display text. The required RAM or ROM is also implemented inside the CPLD chip.

9.4 A VHDL Sync Generation Example: UP1core VGA_SYNC

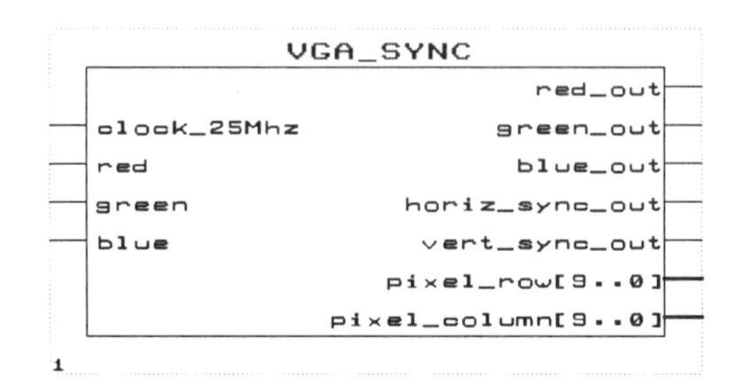

The UP1core function, VGA_SYNC can be used to generate the timing signals needed for a VGA video display. Although VGA_SYNC is written in VHDL, like the other UP1core functions it can be used as a symbol in a design created with any entry method.

The following VHDL code generates the horizontal and vertical sync signals, by using 10-bit counters, H_count for the horizontal count and V_count for the vertical count. H_count and V_count generate a pixel row and column address that is output and available for use by other processes. User logic uses these signals to determine the x and y coordinates of the present video location. The pixel address is used in generating the image's RGB color data. The internal logic uses the onboard 25.175 MHz clock with counters to produce video sync timing signals like those seen in figures 9.3 and 9.4. This process is used in all of the video examples that follow.

```
LIBRARY IEEE;
USE  IEEE.STD_LOGIC_1164.ALL;
USE  IEEE.STD_LOGIC_ARITH.ALL;
USE  IEEE.STD_LOGIC_UNSIGNED.ALL;

ENTITY VGA_SYNC IS
    PORT( clock_25Mhz, red, green, blue  : IN   STD_LOGIC;
          red_out, green_out, blue_out   : OUT  STD_LOGIC;
          horiz_sync_out, vert_sync_out  : OUT  STD_LOGIC;
          pixel_row, pixel_column        : OUT  STD_LOGIC_VECTOR( 9 DOWNTO 0 ));
END VGA_SYNC;

ARCHITECTURE a OF VGA_SYNC IS
    SIGNAL horiz_sync, vert_sync                  : STD_LOGIC;
    SIGNAL video_on, video_on_v, video_on_h : STD_LOGIC;
    SIGNAL h_count, v_count                       : STD_LOGIC_VECTOR( 9 DOWNTO 0 );

BEGIN

                                    -- video_on is High only when RGB data is displayed
video_on <= video_on_H AND video_on_V;
```

```
PROCESS
   BEGIN
   WAIT UNTIL( clock_25Mhz'EVENT ) AND ( clock_25Mhz = '1' );

                    --Generate Horizontal and Vertical Timing Signals for Video Signal
                    -- H_count counts pixels (640 + extra time for sync signals)
                    --
                    -- Horiz_sync  ------------------------------------------__________--------
                    -- H_count      0          640                  659   755   799
                    --
   IF ( h_count = 799 ) THEN
      h_count <= "0000000000";
   ELSE
      h_count <= h_count + 1;
   END IF;

                    --Generate Horizontal Sync Signal using H_count
   IF ( h_count <= 755 ) AND  (h_count => 659 ) THEN
      horiz_sync <= '0';
   ELSE
      horiz_sync <= '1';
   END IF;
                    --V_count counts rows of pixels (480 + extra time for sync signals)
                    --
                    -- Vert_sync     ----------------------------------------________------------
                    -- V_count       0             480           493-494        524
                    --
   IF ( v_count >= 524 ) AND ( h_count => 699 ) THEN
      v_count <= "0000000000";
   ELSIF ( h_count = 699 ) THEN
      v_count <= v_count + 1;
   END IF;
                    -- Generate Vertical Sync Signal using V_count
   IF ( v_count <= 494 ) AND ( v_count = >493 ) THEN
      vert_sync <= '0';
   ELSE
      vert_sync <= '1';
   END IF;
                    -- Generate Video on Screen Signals for Pixel Data
   IF ( h_count <= 639 ) THEN
      video_on_h   <= '1';
      pixel_column <= h_count;
   ELSE
      video_on_h <= '0';
   END IF;

   IF ( v_count <= 479 ) THEN
      video_on_v <= '1';
      pixel_row    <= v_count;
   ELSE
      video_on_v <= '0';
   END IF;
```

```
                            -- Put all video signals through DFFs to eliminate
                            -- any delays that can cause a blurry image
                            -- Turn off RGB outputs when outside video display area
    red_out         <= red AND video_on;
    green_out       <= green AND video_on;
    blue_out        <= blue AND video_on;
    horiz_sync_out  <= horiz_sync;
    vert_sync_out   <= vert_sync;

    END PROCESS;
END a;
```

To turn off RGB data when the pixels are not being displayed the video_on signals are generated. Video_on is gated with the RGB inputs to produce the RGB outputs. Video_on is low during the time that the beam is resetting to the start of a new line or screen. They are used in the logic for the final RGB outputs to force them to the zero state. VGA_SYNC also puts the all of video outputs through a final register to eliminate any timing differences in the video outputs. VGA_SYNC outputs the pixel row and column address.

9.5 Final Output Register for Video Signals

The final video output for the RGB and sync signals in any design should be directly from a flip-flop output. Even a small time delay of a few nanoseconds from the logic that generates the RGB color signals will cause a blurry video image. Since the RGB signals must be delayed a 25Mhz clock period to eliminate any possible timing delays, the sync signals must also be delayed by clocking them through a D flip-flop. If the outputs all come directly from a flip-flop output, the video signals will all change at the same time and a sharper video image is produced. The last few lines of VHDL code in the UP1core VGA_SYNC design generate this final output register.

9.6 Required Pin Assignments for Video Output

The UP 1 board requires the following FLEX chip pins be defined in the project's *.gdf, *.acf file, or elsewhere in your design in order to display the video signals:

```
Clock        : INPUT_PIN = 91;       -- 25Mhz Clock Input Pin
Red          : OUTPUT_PIN = 236;     -- Red Data Signal Output Pin
Blue         : OUTPUT_PIN = 238;     -- Blue Data Signal Output Pin
Green        : OUTPUT_PIN = 237;     -- Green Data Signal Output Pin
Horiz_Sync   : OUTPUT_PIN = 240;     -- Horizontal Sync Signal Output Pin
Vert_Sync    : OUTPUT_PIN = 239;     -- Vertical Sync Signal Output Pin
```

These pins are hard wired on the UP 1 board to the VGA connector and cannot be changed.

9.7 Video Examples

As a simple video example using the VGA_SYNC function, the following schematic is a video simulation of a red LED. When the PB1 pushbutton is hit, the color of the entire video screen will change from black to red.

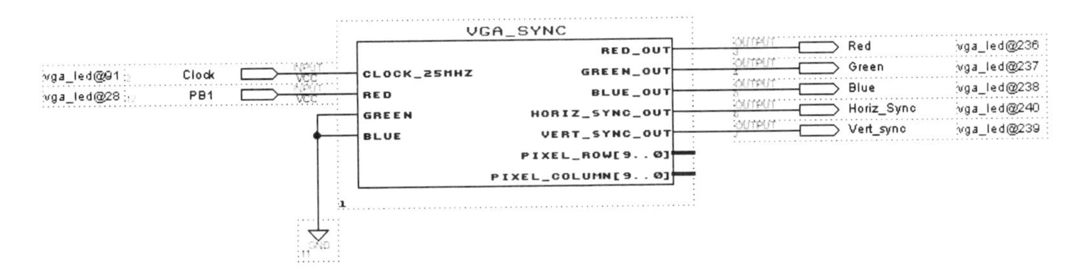

VGA_SYNC outputs the pixel row and column address. Pixel_row and Pixel_column are normally inputs to user logic that in turn generates the RGB color data. Here is a simple example that uses the pixel_column output to generate the RGB inputs. Bits 7, 6, and 5 of the pixel_column count are connected to the RGB data. Since bits 4 through 0 of pixel column are not connected, RGB color data will only change once every 32 pixels across the screen. This in turn generates a sequence of color bars in the video output. The color bars display the eight different colors that can be generated by the three digital RGB outputs.

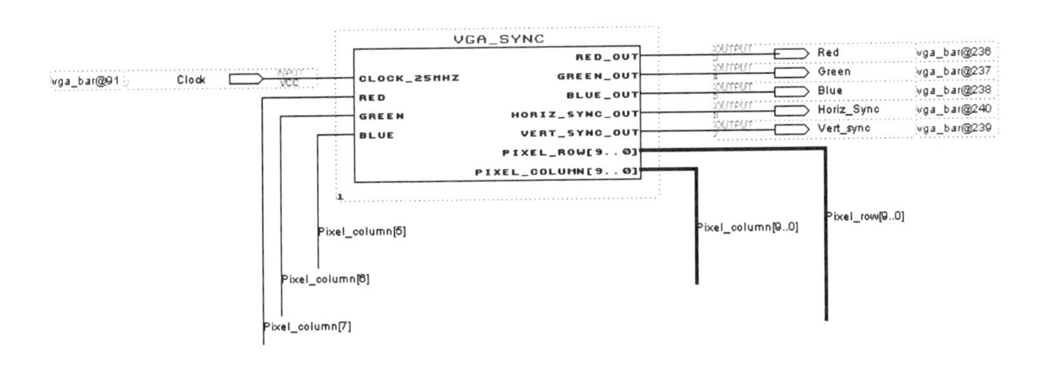

9.8 A Character Based Video Design

One option is a video display that contains mainly textual data. For this approach, a pixel pattern or font is needed to display each different character. The character font can be stored in a ROM implemented inside the FLEX CPLD. A memory initialization file, *.mif, can be used to initialize the ROM contents during download. Given the memory limitations inside the FLEX CPLD, one option that fits is a display of 40 characters by 30 lines.

Each letter, number, or symbol is a pixel image from the 8 by 8 character font. To make the characters larger, each dot in the font maps to a 2 by 2 pixel block so that a single character requires 16 by 16 pixels. This was done by dividing the row and column counters by 2. Recall that in binary, division by powers of two can be accomplished by truncating the lower bits, so no hardware is needed for this step. The row and column counters provide inputs to circuits that address the character font ROM and determine the color of each pixel. The clock used is the onboard 25.175MHz clock and other timing signals needed are obtained by dividing this clock down in hardware.

9.9 Character Selection and Fonts

Because the screen is constantly being refreshed and the video image is being generated on-the-fly as the beam moves across the video display, it is necessary to use other registers, ROM, or RAM inside the CPLD to hold and select the characters to be displayed on the screen. Each location in this character ROM or RAM contains only the starting address of the character font in font ROM. Using two levels of memory results in a design that is more compact and uses far less memory bits. This technique was used on early generation computers before the PC.

Here is an example implementation of a character font used in the UP1core function, char_ROM. To display an "A" the character ROM would contain only the starting address 000001 for the font table for "A". The 8 by 8 font in the character generation ROM would generate the letter "A" using the following eight memory words:

Address	Font Data
000001000 :	00011000 ;
000001001 :	00111100 ;
000001010 :	01100110 ;
000001011 :	01111110 ;
000001100 :	01100110 ;
000001101 :	01100110 ;
000001110 :	01100110 ;
000001111 :	00000000 ;

Figure 9.6 Font Memory Data for the Character "A".

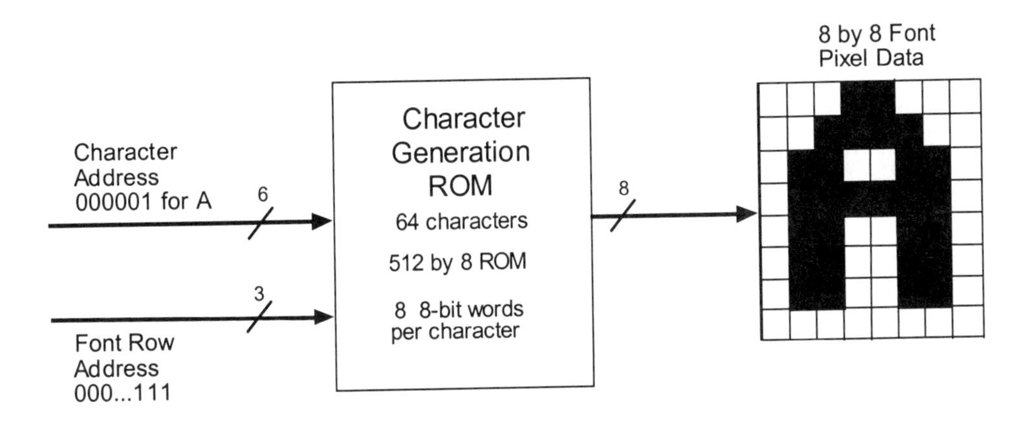

Figure 9.7 Accessing a Character Font Using a ROM.

The column counters are used to select each font bit from left to right in each word of font memory as the video signal moves across a row. This value is used to drive the logic for the RGB signals so that a "0" font bit has a different color from a "1". Using the low three character font row address bits, the row counter would select the next memory location from the character font ROM when the display moves to the next row.

A 3-bit font column address can be used with a multiplexer to select the appropriate bit from the ROM output word to drive the RGB pixel color data. The Character font ROM and the multiplexer are contained in the UP1core char_ROM as shown below. The VHDL code declares the memory size using the LPM_ROM function and the tcgrom.mif file contains the initial values or font data for the ROM.

CHAR_ROM

character_address[5..0]
font_row[2..0]
font_col[2..0]
rom_mux_output

1

```
LIBRARY IEEE;
USE  IEEE.STD_LOGIC_1164.ALL;
USE  IEEE.STD_LOGIC_ARITH.ALL;
USE  IEEE.STD_LOGIC_UNSIGNED.ALL;
LIBRARY lpm;
USE lpm.lpm_components.ALL;
```

```
ENTITY Char_ROM IS
  PORT(   character_address     : IN     STD_LOGIC_VECTOR( 5 DOWNTO 0 );
          font_row, font_col    : IN     STD_LOGIC_VECTOR( 2 DOWNTO 0 );
          rom_mux_output        : OUT    STD_LOGIC);
END Char_ROM;

ARCHITECTURE a OF Char_ROM IS
  SIGNAL  rom_data              : STD_LOGIC_VECTOR( 7 DOWNTO 0 );
  SIGNAL  rom_address           : STD_LOGIC_VECTOR( 8 DOWNTO 0 );
BEGIN
                          -- Small 8 by 8 Character Generator ROM for Video Display
                          -- Each character is 8 8-bit words of pixel data
 char_gen_rom: lpm_rom
   GENERIC MAP (
       lpm_widthad           => 9,
       lpm_numwords          => 512,
       lpm_outdata           => "UNREGISTERED",
       lpm_address_control   => "UNREGISTERED",
                             -- Reads in mif file for character generator font data
       lpm_file              => "tcgrom.mif",
       lpm_width             => 8)
   PORT MAP ( address => rom_address, q = > rom_data);
     rom_address <= character_address & font_row;
                          -- Mux to pick off correct rom data bit from 8-bit word
                          -- for on screen character generation
     rom_mux_output <= rom_data (
           (CONV_INTEGER( NOT font_col( 2 DOWNTO 0 ))) );
END a;
```

Table 9.1 Character Address Map for 8 by 8 Font ROM.

CHAR	ADDRESS	CHAR	ADDRESS	CHAR	ADDRESS	CHAR	ADDRESS
@	*00*	P	*20*	Space	*40*	0	*60*
A	*01*	Q	*21*	!	*41*	1	*61*
B	*02*	R	*22*	"	*42*	2	*62*
C	*03*	S	*23*	#	*43*	3	*63*
D	*04*	T	*24*	$	*44*	4	*64*
E	*05*	U	*25*	%	*45*	5	*65*
F	*06*	V	*26*	&	*46*	6	*66*
G	*07*	W	*27*	'	*47*	7	*67*
H	*10*	X	*30*	(	*50*	8	*70*
I	*11*	Y	*31*	)	*51*	9	*71*
J	*12*	Z	*32*	*	*52*	A	*72*
K	*13*	[	*33*	+	*53*	B	*73*
L	*14*	Dn Arrow	*34*	,	*54*	C	*74*
M	*15*	]	*35*	-	*55*	D	*75*
N	*16*	Up Arrow	*36*	.	*56*	E	*76*
O	*17*	Lft Arrow	*37*	/	*57*	F	*77*

A 16 by 16 pixel area is used to display a single character with the character font. As the display moves to another character outside of the 16 by 16 pixel area, a different location is selected in the character RAM using the high bits of the row and column counters. This in turn selects another location in the character font ROM to display another character.

Due to limited ROM space, only the capital letters, numbers and some symbols are provided. Table 9.1 shows the alphanumeric characters followed by the high six bits of its octal character address in the font ROM. For example, a space is represented by octal code 40. The repeated letters A-F were used to simplify the conversion and display of hexadecimal values.

9.10 VHDL Character Display Design Examples

The UP1cores VGA_SYNC and CHAR_ROM are designed to be used together to generate a text display. CHAR_ROM contains an 8 by 8 pixel character font. In the following schematic, a test pattern with 40 characters across with 30 lines down is displayed. Examining the RGB inputs on the VGA_SYNC core you can see that characters will be white (111 = RGB) with a red (100 = RGB) background. Each character uses a 16 by 16 pixel area in the 640 by 480 display area. Since the low bit in the pixel row and column address is skipped in the font row and font column ROM inputs, each data bit from the font is a displayed in a 2 by 2 pixel area. Since pixel row bits 9 to 4 are used for the character address a new character will be displayed every 16th pixel row or character line. Division by 16 occurs without any logic since the low four bits are not connected.

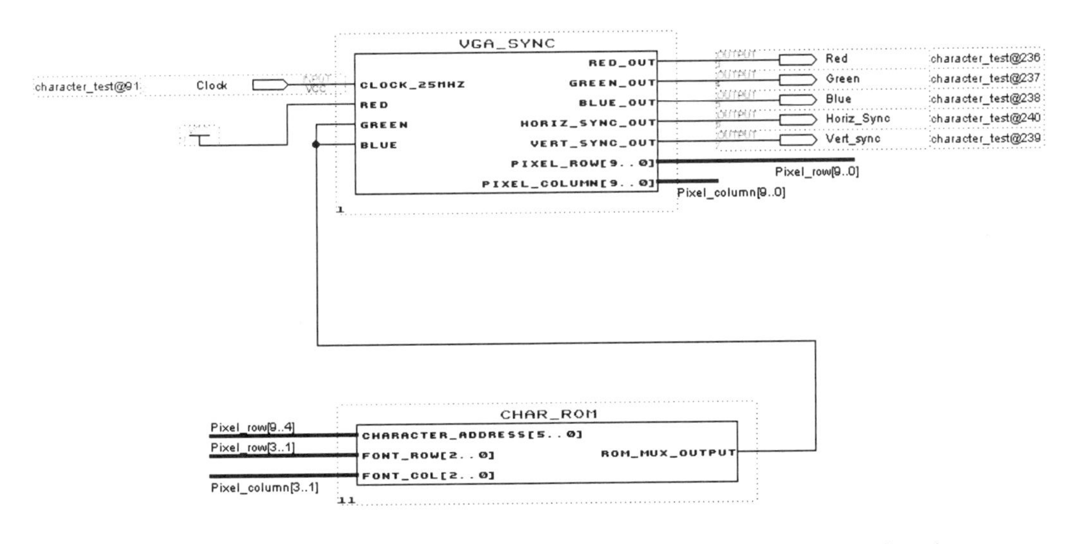

Normally, more complex user designed logic is used to generate the character address. The video example shown in Figure 9.8 is an implementation of the MIPS RISC processor core. The values of major busses are displayed in hexadecimal and it is possible to single step through instructions and watch the values on the video display. This example includes both constant and variable character display areas. The video setup is the same as the schematic, but additional logic is used to generate the character address.

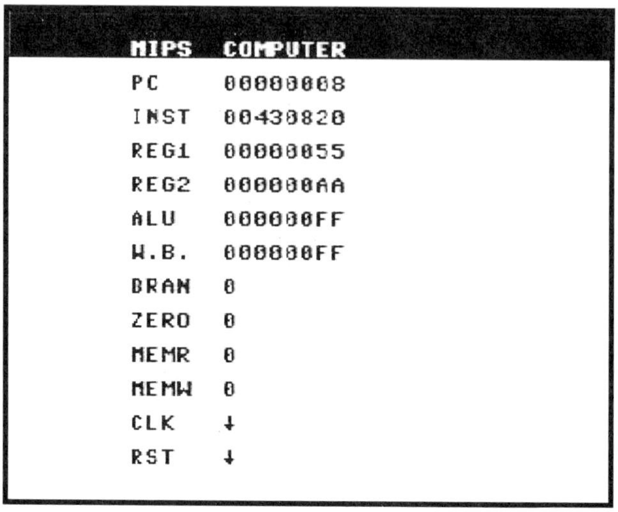

```
MIPS  COMPUTER
PC    00000008
INST  00430820
REG1  00000055
REG2  000000AA
ALU   000000FF
W.B.  000000FF
BRAN  0
ZERO  0
MEMR  0
MEMW  0
CLK   ↓
RST   ↓
```

Figure 9.8 MIPS Computer Video Output.

Pixel row address and column address counters are used to determine the current character column and line position on the screen. They are generated as the image scans across the screen with the VGA_SYNC core by using the high six bits of the pixel row and pixel column outputs. Each character is a 16 by 16 block of pixels. The divide by 16 operation just requires truncation of the low four bits of the pixel row and column. The display area is 40 characters by 30 lines.

Constant character data for titles in the left column is stored in a small ROM called the character format ROM. This section of code sets up the format ROM that contains the character addresses for the constant character data in the left column of the video image for the display.

```
                                    -- Character Format ROM for Video Display
                                    -- Displays constant format character data
                                    -- on left side of Display area
format_rom: lpm_rom
  GENERIC MAP (
    lpm_widthad            => 6,
    lpm_numwords           =>60,
    lpm_outdata            => "UNREGISTERED",
    lpm_address_control    => "UNREGISTERED",
                                    -- Reads in mif file for data display titles
    lpm_file               =>"format.mif",
    lpm_width              => 6)
```

Each 25Mhz clock cycle, a process containing a series of nested CASE statements is used to select the character to display as the image scans across the screen. The CASE statements check the row and column counter outputs from the sync unit to determine the exact character column and character line that is currently being displayed. The CASE statements then output the character address for the desired character to the char_ROM UP1core.

Table 9.1 lists the address of each character in the font ROM. Alphabetic characters start at octal location 01 and numbers start at octal location 60. Octal location 40 contains a space that is used whenever no character is displayed. When the display is in the left column, data from the format_ROM is used. Any unused character display areas must select the space character that has blank or all zero font data.

Hexadecimal variables in the right column in Figure 9.8 are generated by using 4-bit data values from the design to index into the character font ROM. As an example, the value "11" & PC(7 DOWNTO 4), when used as the character address to the UP1core, char_ROM, will map into the character font for 0..9 and A..F. The actual hex character selected is based on the current value of the 4 bits in the VHDL signal, PC. As seen in the last column of Table 9.1, the letters, A..F, appear again after decimal numbers in the font ROM to simplify this hexadecimal mapping conversion.

9.11 A Graphics Memory Design Example

For another example, assume the display will be used to display only graphics data. The FLEX 10K20 EABs contain 12K bits of memory. If only two colors are used in the RGB signals, one bit will be required for each pixel in the video RAM. If a 64 by 64 pixel video RAM was implemented in the FLEX chip it would use 4K bits of the chip's 12K-bit memory. For full color RGB data of three bits per pixel, a 64 by 64 pixel RAM would use all of the 12K on-chip memory and no memory would be left for the remainder of the design.

Pixel memory must always be in read mode whenever RGB data is displayed. To avoid flicker and memory access conflicts, designs should update pixel RAM and other signals that produce the RGB output, during the time the RGB data is not being displayed.

When the scan of each horizontal line is complete there are around 160 clock cycles before the next RGB value is needed, as seen in Figure 9.9.

In most cases, calculations that change the video image should be performed during this off-screen period of time to avoid memory conflicts with the readout of video RAM or other registers which are used to produce the RGB video pixel color signals. Since pixel memory is limited, complex graphic designs with higher resolutions will require another approach.

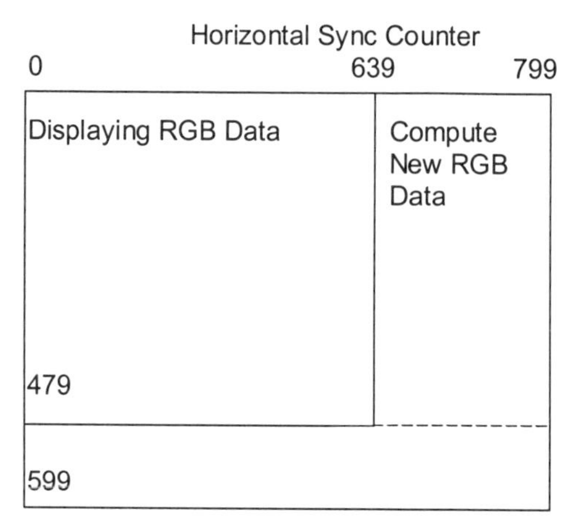

Figure 9.9 Display and Compute clock cycles available in a single Video Frame.

9.12 Video Data Compression

Here are some ideas to save memory and produce more complex graphics. Compress the video pixel data in memory and uncompress it on-the-fly as the video signal is generated. One compression technique that works well is run length encoding (RLE). The RLE compression technique only requires a simple state machine and a counter for decoding.

In RLE, the pixels in the display are encoded into a sequence of length and color fields. The length field specifies the number of sequentially scanned pixels with the same color. In simple color images, substantial data compression can be achieved with RLE and it is used in PCs to encode color bitmaps. In our examples, Matlab was used to read bitmaps into a two-dimensional array and then a Matlab program was written to output an RLE encoded version directly to a *.mif file. This program is available on the CDROM. Bitmap file formats and some C utilities to help read bitmaps can be found on the web.

Many early video games, such as Pong, have a background color with a few moving images. In such cases, the background image can be the default color value and not stored in video RAM. Hardware comparators can check the row and column counts as the video signal is generated and detect when another image other than the background should be displayed. When the comparator signals that the row and column count matches the image location, the image's color data instead of the normal background data is switched into the RGB output using gates or a multiplexer.

The image can be made to move if its current row and column location is stored in registers and the output of these registers are used as the comparator input. Additional logic can be used to increment or decrement the image's location registers slowly over time and produce motion. Multiple hardware comparators can be used to support several fixed and moving images. These moving images are also called sprites. This approach was used in early-generation video games.

9.13 Video Color Mixing using Dithering

Although the hardware supports only eight different pixel colors, there is a technique that can be used to generate up to twenty-seven unique colors. The screen is refreshed at 60Hz, but flicker is almost undetected by the human eye at 30Hz. So, in odd refresh scans one pixel color is used and in even refresh scans another pixel color is used. This 30Hz color mixing or dithering technique works best if large areas always have both colors arranged in a checkerboard pattern. Alternating scans use the inverse checkerboard colors. At 30Hz, the eye can detect color intensity flicker in large regions unless the alternating checkerboard pattern is used. Each of the three RGB colors can have three possible values 0-0, 0-1, and 1-1 in alternating scans. Note that 0-1 and 1-0 will produce the same color. This generates 3^3 or twenty-seven unique colors.

9.14 VHDL Graphics Display Design Example

This simple graphics example will generate a ball that bounces up and down on the screen. As seen in Figure 9.10, the ball is red and the background is white. This example requires the VGA_SYNC design from Section 9.4 to generate the video sync and the pixel address signals. The pixel_row signal is used to determine the current row and the pixel_column signal determines the current column. Using these current row and column addresses, the process RGB_Display generates the red ball on the white background and produces the ball_on signal which displays the red ball using the logic in the red, green, and blue equations. Ball_X_pos and Ball_y_pos are the current address of the center of the ball. Size is the size of the square ball.

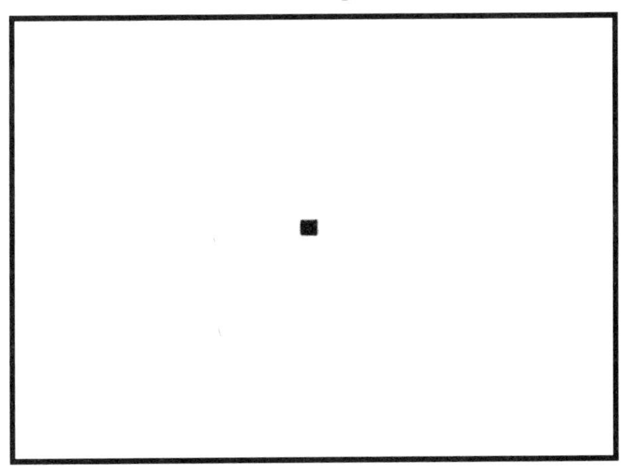

Figure 9.10 Bouncing Ball Video Output.

The process Move_Ball moves the ball a few pixels every vertical sync and checks for bounces off of the walls. Ball_motion is the number of pixels to move the ball at each vertical sync clock. The VGA_SYNC process is also used to generate sync signals and pixel addresses but is not shown in the code below.

```
ENTITY ball IS
    PORT(
        SIGNAL Red, Green, Blue          : OUT STD_LOGIC;
        SIGNAL vert_sync_out             : IN STD_LOGIC;
        SIGNAL pixel_row, pixel_column   : IN STD_LOGIC_VECTOR( 9 DOWNTO 0 ));
END ball;
ARCHITECTURE behavior OF ball IS
                                        -- Video Display Signals
        SIGNAL reset, Ball_on, Direction : STD_LOGIC;
        SIGNAL Size                      : STD_LOGIC_VECTOR( 9 DOWNTO 0 );
        SIGNAL Ball_Y_motion             : STD_LOGIC_VECTOR( 9 DOWNTO 0 );
        SIGNAL Ball_Y_pos, Ball_X_pos    : STD_LOGIC_VECTOR( 9 DOWNTO 0 );
BEGIN                                   -- Size of Ball
    Size        <= CONV_STD_LOGIC_VECTOR (8,10 );
                                        -- Ball center  X address
    Ball_X_pos <= CONV_STD_LOGIC_VECTOR( 320,10 );
                                        -- Colors for pixel data on video signal
    Red         <= '1';                 -- Turn off Green and Blue to make
                                        -- color Red when displaying ball
    Green       <= NOT Ball_on;
    Blue        <= NOT Ball_on;

RGB_Display:
    PROCESS ( Ball_X_pos, Ball_Y_pos, pixel_column, pixel_row, Size )
    BEGIN
                                        -- Set Ball_on = '1' to display ball
        IF (  '0' & Ball_X_pos   <= pixel_column + Size ) AND
           ( Ball_X_pos + Size >= '0' & pixel_column    ) AND
           ( '0' & Ball_Y_pos    <= pixel_row + Size     ) AND
           ( Ball_Y_pos + Size >= '0' & pixel_row       ) THEN
                Ball_on <= '1';
        ELSE
                Ball_on <= '0';
        END IF;
    END PROCESS RGB_Display;

Move_Ball:
    PROCESS
    BEGIN
                            -- Move ball once every vertical sync
    WAIT UNTIL Vert_sync'EVENT AND Vert_sync = '1';
                            -- Bounce off top or bottom of screen
        IF ('0' & Ball_Y_pos) >= CONV_STD_LOGIC_VECTOR(480,10) - Size THEN
            Ball_Y_motion <= - CONV_STD_LOGIC_VECTOR(2,10);
        ELSIF  Ball_Y_pos <= Size   THEN
           Ball_Y_motion <= CONV_STD_LOGIC_VECTOR(2,10);
        END IF;
```

```
                    -- Compute next ball Y position
        Ball_Y_pos <= Ball_Y_pos + Ball_Y_motion;
    END PROCESS Move_Ball;
END behavior;
```

9.15 Laboratory Exercises

1. Design a video output display that displays a large version of your initials. Hint: use the character generation ROM, the Video Sync UP1core, and some of the higher bits of the row and column pixel counters to generate larger characters.

2. Modify the bouncing ball example to bounce and move in both the X and Y directions. You will need to add code for motion in two directions and check additional walls for a bounce condition.

3. Modify the bouncing ball example to move up or down based on input from the two pushbuttons.

4. Modify the example to support different speeds. Read the speed of the ball from the FLEX DIP switches.

5. Draw a more detailed ball in the bouncing ball example. Use a small ROM to hold a small detailed color image of a ball.

6. Make a Pong-type video game by using pushbutton input to move a paddle up and down that the ball will bounce off of.

7. Design your own video game with graphics. Some ideas include breakout, space invaders, Tetris, a slot machine, poker, craps, blackjack, pinball, and roulette. Keep the graphics simple so that the design will fit on the FLEX chip. If the video game needs a random number generator, information on random number generation can be found in Appendix A.

8. Use the character font ROM and the ideas from the MIPS character output example to add video character output to another complex design.

9. Using Matlab or C, write a program to convert a color bitmap into a *.mif file with run-length encoding. Design a state machine to read out the memory and generate the RGB color signals to display the bitmap. Use a reduced resolution pixel size such as 160 by 120. Find a bitmap to display or create one with a paint program. It will work best if the bitmap is already 160 by 120 pixels or smaller. A school mascot or your favorite cartoon character might make an interesting choice. 12K bits of memory are available in the FLEX 10K20 so a 12-bit RLE format with nine bits for length and three bits for color can be used with up to 1024 locations. This means that the bitmap can only have 1024 color changes as the image is scanned across the display. Simple images such as cartoons have fewer color changes. A Matlab example is on the CDROM.

10. Add color mixing or dithering with 27 colors to the previous problem. The 3-bit color code in the RLE encoded memory can be used to map into a 6-bit color palette. The color palette contains two 3-bit RGB values used for color mixing or interlacing. The color palette memory selects 8 different colors out of 27. The program translating the bitmap should select the 8 closest colors for the color palette.

CHAPTER 10

Communications: Interfacing to the PS/2 Keyboard

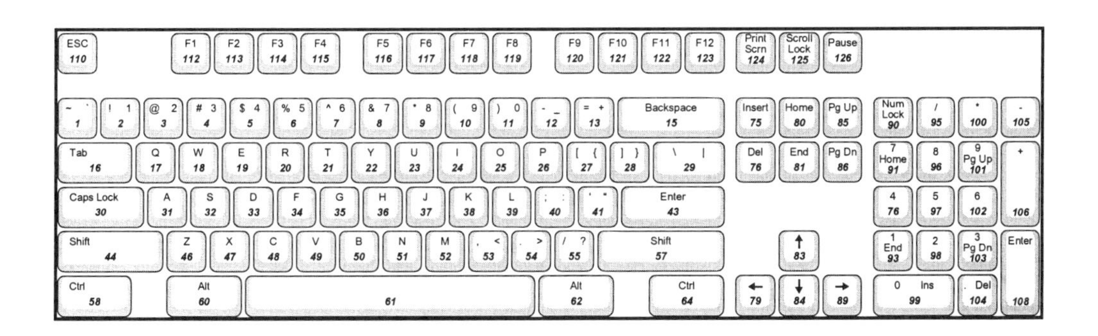

10 Communications: Interfacing to the PS/2 Keyboard

The Altera UP 1 and UP 1X boards support the use of either a mouse or keyboard using a PS/2 connector on the board. This provides only the basic electrical connections from the PS/2 cable and the FLEX chip. It is necessary to design a hardware interface using logic in the FLEX chip to communicate with a keyboard or a mouse. Serial-to-parallel conversion using a shift register is required.

10.1 PS/2 Port Connections

The PS/2 port consists of 6 pins including ground, power (VDD), keyboard data, and a keyboard clock line. The UP 1 board supplies the power to the mouse or keyboard. Two lines are not used. The keyboard data line is pin 31 on the FLEX chip, and the keyboard clock line is pin 30. Pins must be specified in one of the design files. Both the clock and data lines are open collector and bi-directional. The clock line is normally controlled by the keyboard, but it can also be driven by the computer system or in this case the FLEX chip, when it wants to stop data transmissions from the keyboard. Both the keyboard and the system can drive the data line. The data line is the sole source for the data transfer between the computer and keyboard. The keyboard and the system can exchange several commands and messages as seen in Tables 10.1 and 10.2.

Table 10.1 PS/2 Keyboard Commands and Messages.

Commands Sent to Keyboard	Hex Value
Reset Keyboard Keyboard returns AA, 00 after self-test	FF
Resend Message	FE
Set key typematic (autorepeat) XX is scan code for key	FB, XX
Set key make and break	FC, XX
Set key make	FD, XX
Set all key typematic, make and break	FA
Set all keys make	F9
Set all keys make and break	F8
Make all keys typematic (autorepeat)	F7
Set to Default Values	F6
Clear Buffers and start scanning keys	F4
Set typematic (autorepeat) rate and delay Set typematic (autorepeat) rate and delay Bits 6 and 5 are delay (250ms to 1 sec) Bits 4 to 0 are rate (all 0's-30x/sec to all 1's 2x/sec)	F3, XX
Read keyboard ID Keyboard sends FA, 83, AB	F2
Set scan code set XX is 01, 02, or 03	F0, XX
Echo	EE
Set Keyboard LEDs XX is 00000 Scroll, Num, and Caps Lock bits 1 is LED on and 0 is LED off	ED, XX

Table 10.2 PS/2 Commands and messages sent by keyboard.

Messages Sent by Keyboard	Hex Value
Resend Message	FE
Two bad messages in a row	FC
Keyboard Acknowledge Command Sent by Keyboard after each command byte	FA
Response to Echo command	EE
Keyboard passed self-test	AA
Keyboard buffer overflow	00

10.2 Keyboard Scan Codes

Keyboards are normally encoded by placing the key switches in a matrix of rows and columns. All rows and columns are periodically checked by the keyboard encoder or "scanned" at a high rate to find any key state changes. Key data is passed serially to the computer from the keyboard using what is known as a scan code. Each keyboard key has a unique scan code based on the key switch matrix row and column address to identify the key pressed.

There are different varieties of scan codes available to use depending on the type of keyboard used. The PS/2 keyboard has two sets of scan codes. The default scan code set is used upon power on unless the computer system sends a command the keyboard to use an alternate set. The typical PC sends commands to the keyboard on power up and it uses an alternate scan code set. To interface the keyboard to the UP 1 board, it is simpler to use the default scan code set since no initialization commands are required.

10.3 Make and Break Codes

The keyboard scan codes consist of 'Make' and 'Break' codes. One make code is sent every time a key is pressed. When a key is released, a break code is sent. For most keys, the break code is a data stream of F0 followed by the make code for the key. Be aware that when typing, it is common to hit the next key(s) before releasing the first key hit.

Using this configuration, the system can tell whether or not the key has been pressed, and if more than one key is being held down, it can also distinguish which key has been released. One example of this is when a shift key is held down. While it is held down, the '3' key should return the value for the '#' symbol instead of the value for the '3' symbol. Also note that if a key is held down, the make code is continuously sent via the typematic rate until it is released, at which time the break code is sent.

10.4 The PS/2 Serial Data Transmission Protocol

The scan codes are sent serially using 11 bits on the bi-directional data line. When neither the keyboard nor the computer needs to send data, the data line and the clock line are High (inactive).

As seen in Figure 10.1, the transmission of a single key or command consists of the following components:

1. A start bit ('0')
2. 8 data bits containing the key scan code in low to high bit order
3. Odd parity bit such that the eight data bits plus the parity bit are an odd number of ones
4. A stop bit ('1')

The following sequence of events occur during a transmission of a command by the keyboard:

1. The keyboard checks to ensure that both the clock and keyboard lines are inactive. Inactive is indicated by a High state. If both are inactive, the keyboard prepares the 'start' bit by dropping the data line Low.
2. The keyboard then drops the clock line Low for approximately 35us.
3. The keyboard will then clock out the remaining 10 bits at an approximate rate of 70us per clock period. The keyboard drives both the data and clock line.
4. The computer is responsible for recognizing the 'start' bit and for receiving the serial data. The serial data, which is 8 bits, is followed by an odd parity bit and finally a High stop bit. If the keyboard wishes to send more data, it follows the 12th bit immediately with the next 'start' bit.

This pattern repeats until the keyboard is finished sending data at which point the clock and data lines will return to their inactive High state. In Figure 10.1 the keyboard is sending a scan code of 16 for the "1" key and it has a zero parity bit.

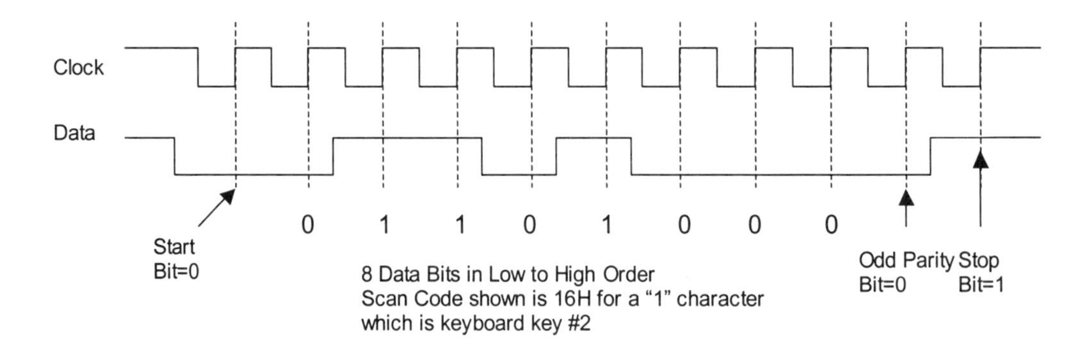

Figure 10.1 Keyboard Transmission of a Scan Code.

When implementing the interface code, it will be necessary to filter the slow keyboard clock to ensure reliable operation with the fast logic inside the FLEX chip. Whenever an electrical pulse is transmitted on a wire, electromagnetic properties of the wire cause the pulse to be distorted and some portions of the pulse may be reflected from the end of the wire. On some PS/2 keyboards and mice there is a reflected pulse on the cable that is strong enough to cause additional clocks to appear on the clock line.

Here is one approach that solves the reflected pulse problem. Feed the PS/2 clock signal into an 8-bit shift register that uses the 25Mhz clock. AND the bits of the shift register together and use the output of the AND gate as the new "filtered" clock. This prevents noise and ringing on the clock line from causing occasional extra clocks during the serial-to-parallel conversion in the FLEX chip.

A few keyboards and mice will work without the clock filter and many will not. They all will work with the clock filter, and it is relatively easy to implement. This circuit is included in the UP1cores for the keyboard and the mouse.

As seen in Figure 10.2, the computer system or FLEX chip sends commands to the PS/2 keyboard as follows:

1. System drives the clock line Low for approximately 60us to inhibit any new keyboard data transmissions. The clock line is bi-directional.

2. System drives the data line Low and then releases the clock line to signal that it has data for the keyboard.

3. The keyboard will generate clock signals in order to clock out the remaining serial bits in the command.

4. The system will send its 8-bit command followed by a parity bit and a stop bit.

5. After the stop bit is driven High, the data line is released.

Upon completion of each command byte, the keyboard will send an acknowledge (ACK) signal, FA, if it received the data successfully. If the system does not release the data line, the keyboard will continue to generate the clock, and upon completion, it will send a 're-send command' signal, FE or FC, to the system. A parity error or missing stop bit will also generate a re-send command signal.

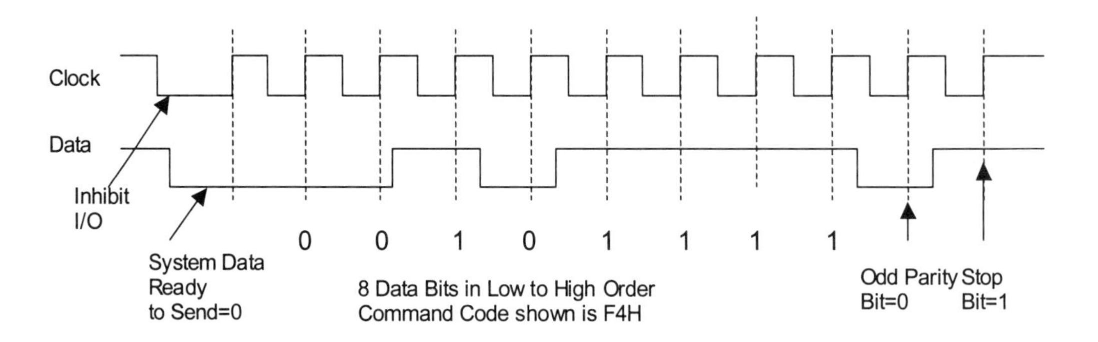

Figure 10.2 System Transmission of a Command to PS/2 Device.

10.5 Scan Code Set 2 for the PS/2 Keyboard

PS/2 keyboards are available in several languages with different characters printed on the keys. A two-step process is required to find the scan code. A key number is used to lookup the scan code. Key numbers needed for the scan code table are shown in Figure 10.3 for the English language keyboard layout.

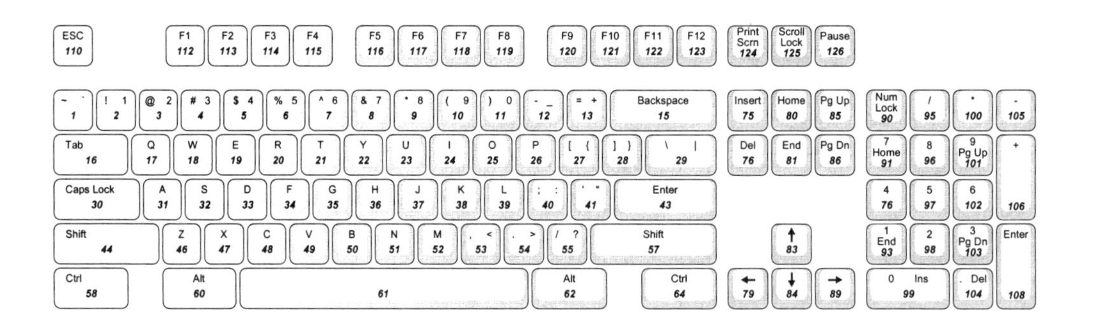

Figure 10.3 Key Numbers for Scan Code.

Each key sends out a make code when hit and a break code when released. When several keys are hit at the same time, several make codes will be sent before a break code.

The keyboard powers up using this scan code as the default. Commands must be sent to the keyboard to use other scan code sets. The PC sends out an initialization command that forces the keyboard to use the other scan code.

The interface is much simpler if the default scan code is used. If the default scan code is used, no commands will need to be sent to the keyboard. The keys in Table 10.3 for the default scan code are typematic (i.e. they automatically repeat the make code if held down).

Table 10.3 Scan Codes for PS/2 Keyboard.

Key#	Make Code	Break Code	Key#	Make Code	Break Code	Key#	Make Code	Break Code
1	0E	F0 0E	31	1C	F0 1C	90	77	F0 77
2	16	F0 16	32	1B	F0 1B	91	6C	F0 6C
3	1E	F0 1E	33	23	F0 23	92	6B	F0 6B
4	26	F0 26	34	2B	F0 2B	93	69	F0 69
5	25	F0 25	35	34	F0 34	96	75	F0 75
6	2E	F0 2E	36	33	F0 33	97	73	F0 73
7	36	F0 36	37	3B	F0 3B	98	72	F0 72
8	3D	F0 3D	38	42	F0 42	99	70	F0 70
9	3E	F0 3E	39	4B	F0 4B	100	7C	F0 7C
10	46	F0 46	40	4C	F0 4C	101	7D	F0 7D
11	45	F0 45	41	52	F0 52	102	74	F0 74
12	4E	F0 4E	43	5A	F0 5A	103	7A	F0 7A
13	55	F0 55	44	12	F0 12	104	71	F0 71
15	66	F0 66	46	1A	F0 1A	105	7B	F0 7B
16	0D	F0 0D	47	22	F0 22	106	79	F0 79
17	15	F0 15	48	21	F0 21	110	76	F0 76
18	1D	F0 1D	49	2A	F0 2A	112	05	F0 05
19	24	F0 24	50	32	F0 32	113	06	F0 06
20	2D	F0 2P	51	31	F0 31	114	04	F0 04
21	2C	F0 2C	52	3A	F0 3A	115	0c	F0 0C
22	35	F0 35	53	41	F0 41	116	03	F0 03
23	3C	F0 3C	54	49	F0 49	117	0B	F0 0B
24	43	F0 43	55	4A	F0 4A	118	83	F0 83
25	44	F0 44	57	59	F0 59	119	0A	F0 0A
26	4D	F0 4D	58	14	F0 14	120	01	F0 01
27	54	F0 54	60	11	F0 11	121	09	F0 09
28	5B	F0 5B	61	29	F0 29	122	78	F0 78
29	5D	F0 5D	62	E0 11	E0 F0 11	123	07	F0 07

The remaining key codes are a function of the shift, control, alt, or num-lock keys.

Table 10.3 (Continued) - Scan Codes for PS/2 Keyboard.

Key	No Shift or Num Lock		Shift*		Num Lock On	
#	Make	Break	Make	Break	Make	Break
76	E0 70	E0 F0 70	E0 F0 12 E0 70	E0 F0 70 E0 12	E0 12 E0 70	E0 F0 70 E0 F0 12
76	E0 71	E0 F0 71	E0 F0 12 E0 71	E0 F0 71 E0 12	E0 12 E0 71	E0 F0 71 E0 T0 12
79	E0 6B	E0 F0 6B	E0 F0 12 E0 6B	E0 F0 6B E0 12	E0 12 E0 6B	E0 F0 6B E0 F0 12
80	E0 6C	E0 F0 6C	E0 F0 12 E0 6C	E0 F0 6C E0 12	E0 12 E0 6C	E0 F0 6C E0 F0 12
81	E0 69	E0 F0 69	E0 F0 12 E0 69	E0 F0 69 E0 12	E0 12 E0 69	E0 F0 69 E0 F0 12
83	E0 75	E0 F0 75	E0 F0 12 E0 75	E0 F0 75 E0 12	E0 12 E0 75	E0 F0 75 E0 F0 12
84	E0 72	E0 F0 72	E0 F0 12 E0 72	E0 F0 72 E0 12	E0 12 E0 72	E0 F0 72 E0 F0 12
85	E0 7D	E0 F0 7D	E0 F0 12 E0 7D	E0 F0 7D E0 12	E0 12 E0 7D	E0 F0 7D E0 F0 12
86	E0 7A	E0 F0 7A	E0 F0 12 E0 7A	E0 F0 7A E0 12	E0 12 E0 7A	E0 F0 7A E0 F0 12
89	E0 74	E0 F0 74	E0 F0 12 E0 74	E0 F0 74 E0 12	E0 12 E0 74	E0 F0 74 E0 F0 12

* When the left Shift Key is held down, the 12 - FO 12 shift make and break is sent with the other scan codes. When the right Shift Key is held down, 59 – FO 59 is sent.

Key	Scan Code		Shift Case *	
#	Make	Break	Make	Break
95	E0 4A	E0 F0 4A	E0 F0 12 E0 4A	E0 12 F0 4A

* When the left Shift Key is held down, the 12 - FO 12 shift make and break is sent with the other scan codes. When the right Shift Key is held down, 59 - FO 59 is sent. When both Shift Keys are down, both sets of codes are sent with the other scan codes.

Key	Scan Code		Control Case, Shift Case		Alt Case	
#	Make	Break	Make	Break	Make	Break
124	E0 12 E0 7C	E0 F0 7C E0 F0 I2	E0 7C	E0 F0 7C	84	F0 84

Key #	Make Code	Control Key Pressed
126 *	EI 14 77 EI F0 14 F0 77	E0 7E E0 F0 7E

* This key does not repeat

10.6 The Keyboard UP1core

The following VHDL code for the keyboard UP1core shown in Figure 10.4 reads the scan code bytes from the keyboard. In this example code, no command is ever sent to the keyboard, so clock and data are always used as inputs and the keyboard power-on defaults are used. To send commands, a more complex bi-directional tri-state clock and data interface is required. The details of such an interface are explained in the next chapter on the PS/2 mouse. The keyboard powers up and sends the self-test code AA and 00 to the FLEX chip before it is downloaded.

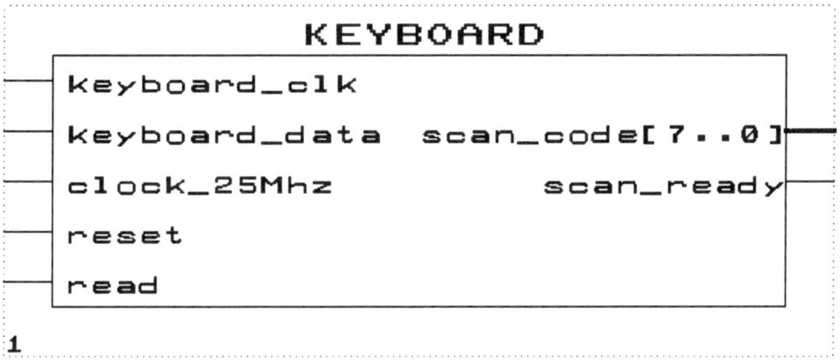

Figure 10.4 Keyboard UP1core

```
LIBRARY IEEE;
USE IEEE.STD_LOGIC_1164.ALL;
USE IEEE.STD_LOGIC_ARITH.ALL;
USE IEEE.STD_LOGIC_UNSIGNED.ALL;

ENTITY keyboard IS
    PORT( keyboard_clk, keyboard_data, clock_25Mhz ,
          reset, read          : IN   STD_LOGIC;
          scan_code            : OUT  STD_LOGIC_VECTOR( 7 DOWNTO 0 );
          scan_ready           : OUT  STD_LOGIC);
END keyboard;

ARCHITECTURE a OF keyboard IS
    SIGNAL INCNT                    : STD_LOGIC_VECTOR( 3 DOWNTO 0 );
    SIGNAL SHIFTIN                  : STD_LOGIC_VECTOR( 8 DOWNTO 0 );
    SIGNAL READ_CHAR                : STD_LOGIC;
    SIGNAL INFLAG, ready_set        : STD_LOGIC;
    SIGNAL keyboard_clk_filtered    : STD_LOGIC;
    SIGNAL filter                   : STD_LOGIC_VECTOR( 7 DOWNTO 0 );

BEGIN
    PROCESS ( read, ready_set )
    BEGIN
        IF read = '1' THEN
            scan_ready <= '0';
        ELSIF ready_set'EVENT AND ready_set = '1' THEN
            scan_ready <= '1';
        END IF;
    END PROCESS;
                        --This process filters the raw clock signal coming from the
                        -- keyboard using a shift register and two AND gates
Clock_filter:
    PROCESS
        BEGIN
```

```
      WAIT UNTIL clock_25Mhz'EVENT AND clock_25Mhz = '1';
      filter ( 6 DOWNTO 0 ) <= filter( 7 DOWNTO 1 ) ;
      filter( 7 ) <= keyboard_clk;
      IF filter = "11111111" THEN
        keyboard_clk_filtered <= '1';
      ELSIF filter = "00000000" THEN
        keyboard_clk_filtered <= '0';
      END IF;
  END PROCESS Clock_filter;
            --This process reads in serial scan code data coming from the keyboard
  PROCESS
  BEGIN
    WAIT UNTIL (KEYBOARD_CLK_filtered'EVENT AND KEYBOARD_CLK_filtered = '1');
    IF RESET = '1' THEN
        INCNT  <= "0000";
        READ_CHAR   <= '0';
    ELSE
        IF KEYBOARD_DATA = '0' AND READ_CHAR = '0' THEN
            READ_CHAR   <= '1';
            ready_set   <= '0';
        ELSE
                    -- Shift in next 8 data bits to assemble a scan code
            IF READ_CHAR = '1' THEN
                IF INCNT < "1001" THEN
                    INCNT   <= INCNT + 1;
                    SHIFTIN( 7 DOWNTO 0 ) <= SHIFTIN( 8 DOWNTO 1 );
                    SHIFTIN( 8 )      <= KEYBOARD_DATA;
                    ready_set    <= '0';
                        -- End of scan code character, so set flags and exit loop
                ELSE
                    scan_code  <= SHIFTIN( 7 DOWNTO 0 );
                    READ_CHAR    <='0';
                    ready_set    <= '1';
                    INCNT   <= "0000";
                END IF;
            END IF;
        END IF;
    END IF;
END PROCESS;
END a;
```

The keyboard clock is filtered in the Clock_filter process using an 8-bit shift register and an AND gate to eliminate any reflected pulses, noise, or timing hazards that can be found on some keyboards. The clock signal in this process is the 25Mhz-system clock. The output signal, keyboard_clk_filtered, will only change if the input signal, keyboard_clk, has been High or Low for eight successive 25Mhz clocks or 320ns. This filters out noise and reflected pulses on the keyboard cable that could cause an extra or false clock signal on the fast FLEX chip. This problem has been observed to occur on some PS/2 keyboards and mice and is fixed by the filter routine.

The RECV_KBD process waits for a start bit, converts the next eight serial data bits to parallel, stores the input character in the signal, charin, and sets a flag, scan_ready, to indicate a new character was read. . The scan_ready or input ready flag is a handshake signal needed to ensure that a new scan code is read in and processed only once. Scan_ready is set whenever a new scan code is received. The input signal, read, resets the scan ready handshake signal.

The process using this code to read the key scan code would need to wait until the input ready flag, scan_ready, goes High. This process should then read in the new scan code value, scan_code. Last, read should be forced High and Low to clear the scan_ready handshake signal.

Since the set and reset conditions for scan_ready come from different processes each with different clocks, it is necessary to write a third process to generate the scan_ready handshake signal using the set and reset conditions from the other two processes. Hitting a common key will send a 1-byte make code and a 2-byte break code. This will produce at least three different scan_code values each time a key is hit and released.

A shift register is used with the filtered clock signals to perform the serial to parallel conversion. No command is ever sent the keyboard and it powers up using scan code set 2. Since commands are not sent to the keyboard, in this example clock and data lines are not bi-directional. The parity bit is not checked.

10.7 A Design Example Using the Keyboard UP1core

Here is a simple design using the Keyboard UP1core. The last byte of the scan code will appear in the seven-segment LED display. Read and Scan_ready are not used in this example.

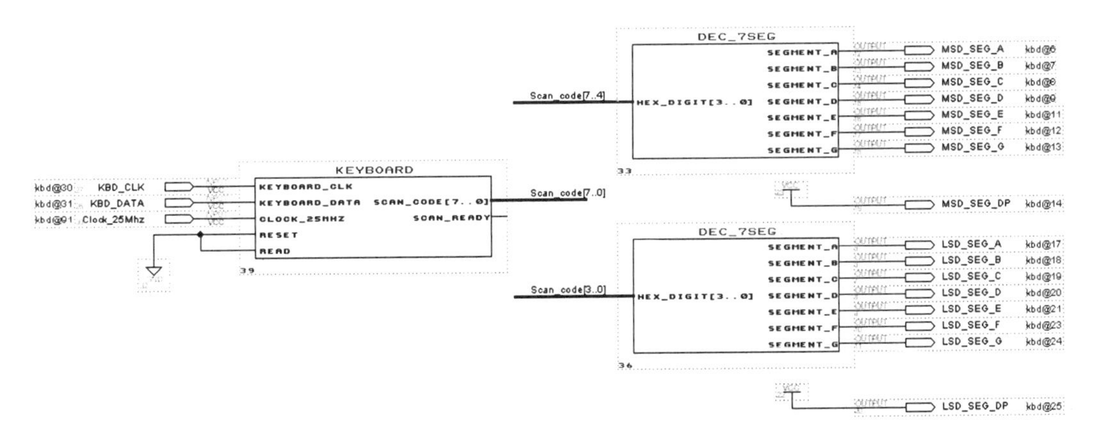

Figure 10.5 Example design using the Keyboard UP1core.

10.8 For Additional Information

The IBM PS/2 Hardware Interface Technical Reference Manual, IBM Corporation, 1988 contains information on the keyboard in the Keyboard and Auxiliary Device Controller Chapter. Scan codes for the alternate scan code set normally used by the PC can be found on the web and in many PC reference manuals.

10.9 Laboratory Exercises

1. Write a VHDL module to read a keyboard scan code and display the entire scan code string in hexadecimal on the VGA display using the VGA_SYNC and CHAR_ROM UP1cores. It will require the use of the read and scan ready handshake lines and a small RAM to hold the scan code bytes.

2. After reading the section on the PS/2 mouse, design an interface that can also send commands to the keyboard. Demonstrate that the design works correctly by changing the status of the keyboard LEDs after reading the new settings from the FLEX DIP switch.

3. Develop a keyboard module that uses the alternate scan code set used by the PC.

4. Write the keyboard module in another HDL such as Verilog.

5. Use the keyboard as a new input device for a video game, the μP1 computer, or another application.

CHAPTER 11

Communications: Interfacing to the PS/2 Mouse

11 Communications: Interfacing to the PS/2 Mouse

Just like the PS/2 keyboard, the PS/2 mouse uses the PS/2 synchronous bi-directional serial communication protocol described in section 10.4 and shown in Figures 10.1 and 10.2. Internally, the mouse contains a ball that rolls two slotted wheels. The wheels are connected to two optical encoders. The two encoders sense x and y motion by counting pulses when the wheels move. It also contains two or three pushbuttons that can be read by the system and a single-chip microcontroller. The microcontroller in the mouse sends data packets to the computer reporting movement and button status.

It is necessary for the computer or in this case the FLEX chip to send the mouse an initialization command to have it start sending mouse data packets. This makes interfacing to the mouse more difficult than interfacing to the keyboard. As seen in Table 11.1, the command value needed for initialization after power up is F4, enable streaming mode.

Table 11.1 PS/2 Mouse Commands.

Commands Sent to Mouse	Hex Value
Reset Mouse	FF
Mouse returns AA, 00 after self-test	
Resend Message	FE
Set to Default Values	F6
Enable Streaming Mode Mouse starts sending data packets at default rate	F4
Disable Streaming Mode	F5
Set sampling rate XX is number of packets per second	F3, XX
Read Device Type	F2
Set Remote Mode	EE
Set Wrap Mode Mouse returns data sent by system	EC
Read Remote Data Mouse sends 1 data packet	EB
Set Stream Mode	EA
Status Request Mouse returns 3-bytes with current settings	E9
Set Resolution XX is 0, 1, 2, 3	E8, XX
Set Scaling 2 to 1	E7
Reset Scaling	E6

Table 11.2 PS/2 Mouse Messages.

Messages Sent by Mouse	Hex Value
Resend Message	FE
Two bad messages in a row	FC
Mouse Acknowledge Command Sent by Mouse after each command byte	FA
Mouse passed self-test	AA

After streaming mode is enabled, the mouse sends data to the system in three byte data packets that contain motion and pushbutton status. The format of a three-byte mouse data packet is seen in Table 11.3.

Table 11.3 PS/2 Mouse Data Packet Format.

	MSB							LSB
Bit	7	6	5	4	3	2	1	0
Byte 1	*Yo*	*Xo*	*Ys*	*Xs*	1	M	R	L
Byte 2	X7	X6	X5	X4	X3	X2	X1	X0
Byte 3	Y7	Y6	Y5	Y4	Y3	Y2	Y1	Y0

L = Left Key Status bit (For buttons 1 = Pressed and 0 = Released)
M = Middle Key Status bit (This bit is reserved in the standard PS/2 mouse protocol, but some three button mice use the bit for middle button status.)
R = Right Key Status bit
X7 – X0 = Moving distance of X in two's complement
(Moving Left = Negative; Moving Right = Positive)
Y7 – Y0 = Moving distance of Y in two's complement
(Moving Up = Positive; Moving Down = Negative)
Xo = X Data Overflow bit (1 = Overflow)
Yo = Y Data Overflow bit (1 = Overflow)
Xs = X Data sign bit (1 = Negative)
Ys = Y Data sign bit (1 = Negative)

11.1 The Mouse UP1core

The UP1core function Mouse is designed to provide a simple interface to the mouse. This function initializes the mouse and then monitors the mouse data transmissions. It outputs a mouse cursor address and button status. The internal operation of the Mouse UP1core is rather complex and the fundamentals are described in the section that follows. Like the other UP1core functions, it is written in VHDL and source code is provided.

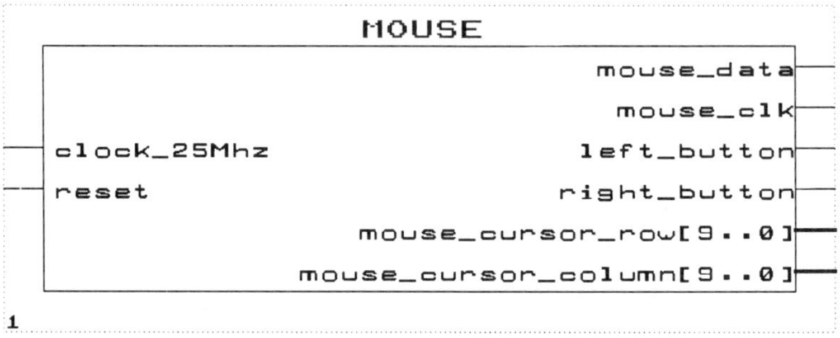

MOUSE

clock_25Mhz
reset
mouse_data
mouse_clk
left_button
right_button
mouse_cursor_row[9..0]
mouse_cursor_column[9..0]
1

To interface to the mouse, a clock filter, serial-to-parallel conversion and parallel-to-serial conversion with two shift registers is required along with a state machine to control the various modes. See Chapter 10 on the PS/2 keyboard for an example of a clock filter design.

11.2 Mouse Initialization

Two lines are used to interface to the mouse, clock and data. The lines must be tri-state bi-directional, since at times they are driven by the mouse and at other times by the FLEX chip. All clock, data, and handshake signals share two tri-state, bi-directional lines, clock and data. These two lines must be declared bi-directional when pin assignments are made and they must have tri-state outputs in the interface. The mouse actually has open collector outputs that can be simulated by using a tri-state output. The mouse always drives the clock signal for any serial data exchanges. The FLEX chip can inhibit mouse transmissions by pulling the clock line Low at any time.

The FLEX chip drives the data line when sending commands to the mouse. When the mouse sends data to the FLEX chip it drives the data line. The tri-state bi-directional handshaking is described in more detail in the IBM PS/2 Technical Reference manual. A simpler version with just the basics for operation with the UP 1 is presented here. Just like the keyboard, the mouse interface is more reliable if a clock filter is used on the clock line.

At power-up, the mouse runs a self-test and sends out the codes AA and 00. The clock and data FLEX chip outputs are tri-stated before downloading the UP 1, so they float High. High turns out to be ready to send for mouse data, so AA and 00 are sent out prior to downloading and need not be considered in the interface. This assumes that the mouse is plugged in before applying power to the UP 1 board and downloading the design.

The default power-up mode is streaming mode disabled. To get the mouse to start sending 3-byte data packets, the streaming mode must be turned on by sending the enable streaming mode command, F4, to the mouse from the FLEX chip. The clock tri-state line is driven Low by the FLEX for at least 60us to inhibit any data transmissions from the mouse. This is the only case when the FLEX chip should ever drive the clock line. The data line is then driven Low by the FLEX chip to signal that the system has a command to send the mouse.

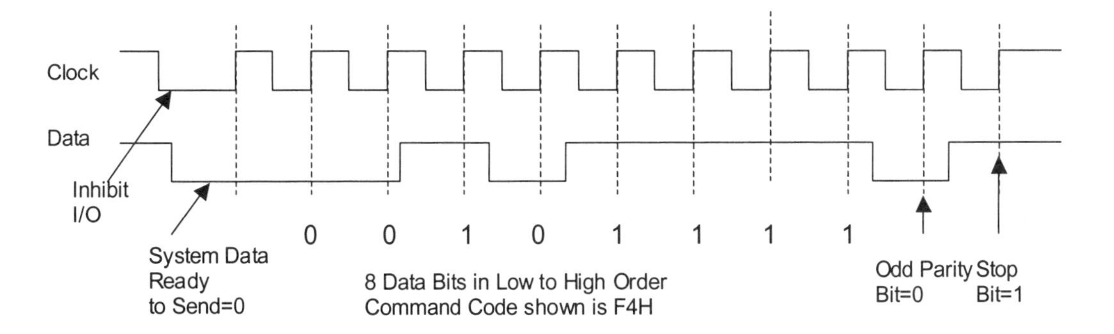

Figure 11.1 Transmission of Mouse Initialization Command.

The clock line is driven High for two clocks at 25Mhz and then tri-stated to simulate an open collector output. This reduces the rise time and reflections on the mouse cable that might be seen by the fast FLEX chip logic as the clock line returns to the High state. As an alternative, the mouse clock input to the FLEX could be briefly disabled while the clock line returns to the High state.

Next the mouse, seeing data Low and clock High, starts clocking in the serial data from the FLEX chip. The data is followed by an odd parity bit and a High stop bit. The handshake signal of the data line starting out Low takes the place of the start bit when sending commands to the mouse.

With the FLEX chip clock and data drivers both tri-stated, the mouse then responds to this message by sending an acknowledge message code, FA, back to the FLEX chip. Data from the mouse includes a Low start bit, eight data bits, an odd parity bit, and a High stop bit. The mouse, as always, drives the clock line for the serial data transmission. The mouse is now initialized.

11.3 Mouse Data Packet Processing

As long as the FLEX chip clock and data drivers remain tri-stated, the mouse then starts sending 3-byte data packets at the power-up default sampling rate of 100 per second. Bytes 2 and 3 of the data packet contain X and Y motion values as was seen in Table 11.2. These values can be positive or negative, and they are in two's complement format.

For a video cursor such as is seen in the PC, the motion value will need to be added to the current value every time a new data packet is received. Assuming 640 by 480 pixel resolution, two 10-bit registers containing the current cursor row and column addresses are needed. These registers are updated every packet by adding the sign extended 8-bit X and Y motion values found in bytes 2 and 3 of the data packet. The cursor normally would be initialized to the center of the video screen at power-up.

11.4 An Example Design Using the Mouse UP1core

In this example design, the mouse drives the seven-segment LED displays. The most-significant display contains the high bits of the cursor row address and the least-significant display contains the high bits of the cursor column address. The left button and the right button drive the MSD and LSD decimal points. The mouse cursor powers up to the center position of the 640 by 480 video screen. Note that the mouse clock and data pins must be bi-directional.

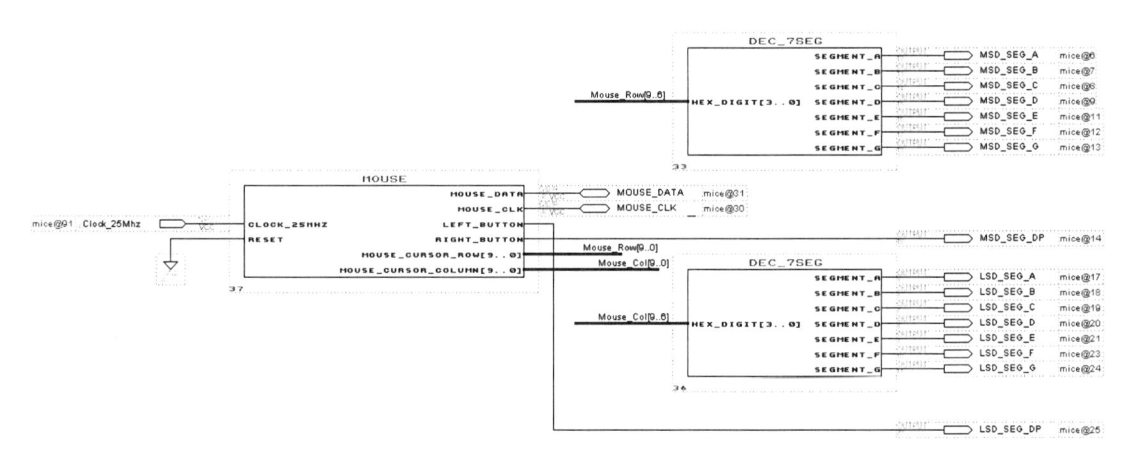

11.5 For Additional Information

The IBM PS/2 Hardware Interface Technical Reference Manual, IBM Corporation, 1988, contains information on the mouse in the Keyboard and Auxiliary Device Controller Chapter.

11.6 Laboratory Exercises

1. Generate a video display that has a moving cursor controlled by the mouse using the Mouse and VGA_Sync UP1cores. Use the mouse buttons to change the color of the cursor.

2. Use the mouse as input to a video etch-a-sketch. Use a monochrome 64 by 64 1-bit pixel RAM in your video design. Display a cursor. To draw a line, the left mouse button should be held down.

3. Use the mouse as an input device in another design with video output or a simple video game such as pong, breakout, or Tetris.

4. Write a mouse driver in Verilog. Use the mouse information provided in sections 11.2 and 11.3.

CHAPTER 12

Robotics: The UP1-bot

Photo: The UP1-bot is a small robot controlled by the UP 1 board

12 Robotics: The UP1-bot

12.1 The UP1-bot Design

The UP1-bot shown in Figure 12.1 is a moving robotics platform designed for the UP 1 board. The UP1-bot is designed to be a small autonomous vehicle that is programmed to move in response to sensory input. A wide variety of sensors can be easily attached to the UP1-bot.

The round platform is cut from plastic and a readily available 7.2V rechargeable battery pack is used to supply power. Two diametrically opposed drive motors move the robot. A third inactive castor wheel or skid is used to provide stability. The robot can move forward, reverse, and rotate in place. Two relatively inexpensive radio control servos are used as drive motors. The UP 1 is programmed to act as the controller. The servos are modified to act as drive motors. The servos are controlled by timing pulses produced by the UP 1 board.

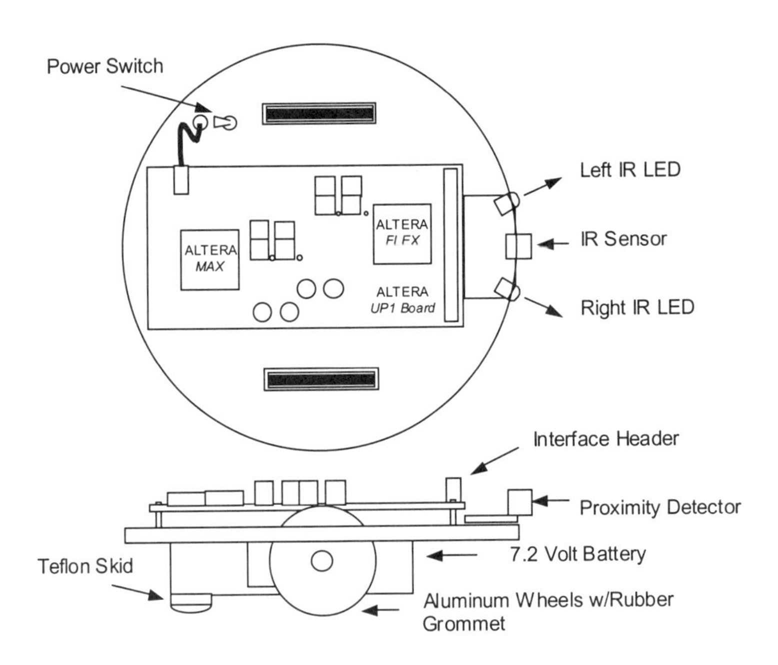

Figure 12.1 The UP1-bot.

12.2 UP1-bot Servo Drive Motors

A typical radio control servo is shown in Figure 12.2. Servos have a drive wheel that is controlled by a coded signal. The servo shown is a Futaba S3003 which is identical, internally, to the Tower TS53J servo. Radio control servomotors are mass-produced for the hobby market and are therefore relatively inexpensive and consistently available. They are ideally suited for robotics applications. Internally, the servo contains a DC drive motor (seen on the left in Figure 12.2), built-in control circuitry, and a gear reduction system. They are small, produce a relatively large amount of torque for their size, and run at the appropriate speed for a robotics drive motor.

Figure 12.2 Left: Radio Control Servo Motor and Right: Servo with Case and Gears Removed.

The control circuitry of the servo uses a potentiometer (variable resistor) that is used to sense the angular position of the output shaft. The potentiometer is the tall component on the right in Figure 12.2. The output shaft of a servo normally travels 180-210 degrees. A single control bit is used to specify the angular position of the shaft. The timing of this bit specifies the angular position for the shaft. The potentiometer senses the angle, and if the shaft is not at the correct angle, the internal control circuit turns the motor in the correct direction until the desired angle is sensed.

The control signal bit specifies the desired angle. The desired angle is encoded using pulse width modulation (PWM). The width of the active high pulse varies from 1-2 ms. A 1ms pulse is 0 degrees, 1.5ms is 90 degrees and a 2 ms pulse is approximately 180 degrees. New timing pulses are sent to the servo every 20 ms.

12.3 Modifying the Servos to make Drive Motors

Normally, a servo has a mechanical stop that prevents it from traveling move than half a revolution. If this stop is removed along with other modifications to the potentiometer, a servo can be converted to a continuously rotating drive motor. Modifications to the servo are not reversible and they will void the warranty.

To modify the servo, open the housing by removing the screws and carefully note the location of the gears, so that they can be reassembled later. The potentiometer can be replaced with two 2.2K ohm ¼ watt resistors or disconnected by cutting the potentiometer shaft shorter and setting it to the center position so that it reports the 90-degree position. A more accurate setting can be achieved by sending the servo a 1.5ms pulse and adjusting the potentiometer until the motor stops moving. The potentiometer can then be glued in place with CA glue. In the center position the potentiometer will have the same resistance from each of the outside pins to the center pin. If the potentiometer is replaced with two resistors, a resistor is connected between each of the two outside pins and the center pin.

In some servos, there will be less mechanical play if the potentiometer is disabled by cutting the center pin and modified by drilling out the stop on the potentiometer so that it can rotate freely. The two resistors are then added to replace the potentiometer in the circuit.

The largest gear in the gear train that drives the output shaft normally has a tab molded on it that serves as the mechanical stop. After removing the screw on the output shaft and removing the large gear, the mechanical stop can be carefully trimmed off with a hobby saw, knife, or small rotary-grinding tool. The servo is then carefully re-assembled.

After modifications, if a pulse shorter than 1.5 ms is sent, the motor will continuously rotate in one direction. If a pulse longer than 1.5 ms is sent the motor will continuously rotate in the other direction. The 1.5 ms or 90-degree position is somctimcs called the neutral position or dead zone. The drive signal to the motor is proportional, so the farther it is from the neutral position the faster it moves. This can be used to control the speed of the motor if the neutral position is carefully adjusted. A pulse width of 0 ms or no pulse will stop the servomotor.

A servo has three wires, +4 to +6 Volt DC power, ground, and the signal wire. The assignment of the three signals on the connector varies among different servo manufacturers. For Futaba servos, the red wire is +5, black is ground, and the white or yellow wire is the pulse width signal line.

On the UP1-bot, the UP 1 board must be programmed to provide the two timing signals to control the servo drive motors.

12.4 VHDL Servo Driver Code for the UP1-bot

To drive the motors a servo signal must be sent every 20 ms with a 0, 1, or 2 ms pulse. The UP 1 board is programmed to produce the timing signals that drive the motors. If no pulse is sent, the motor stops. If a 1 ms pulse is sent, the motor moves clockwise and if a 2 ms pulse is sent the motor moves in the reverse direction, counterclockwise. To move the UP1-bot forward, one motor moves clockwise while the other motor moves counterclockwise. This is because of the way the motors are mounted to the UP1-bot base.

In the code that follows, lmotor_dir and rmotor_dir specify the direction for the left and right motor. If both signals are '1' the UP 1 bot moves forward. The VHDL code actually moves one motor in the opposite direction to move forward. If both are '0' the robot moves in reverse. If one is '1' and the other is '0', the UP 1 bot turns by rotating in place. The two speed controls are lmotor_speed and rmotor_speed. In the speed control signals, '0' is stop and '1' is run. A 1Khz clock is used for the counters in the module. The UP1core function, clk_div, can be used to provide this signal. Two more complex techniques for implementing variable speed control are discussed in problems at the end of the chapter.

```
LIBRARY IEEE;
USE  IEEE.STD_LOGIC_1164.ALL;
USE  IEEE.STD_LOGIC_ARITH.ALL;
USE  IEEE.STD_LOGIC_UNSIGNED.ALL;
ENTITY motor_control IS
     PORT     (clock_1khz                          : IN        STD_LOGIC;
               lmotor_dir, rmotor_dir         : IN STD_LOGIC;
               lmotor_speed, rmotor_speed     : IN STD_LOGIC;
               lmotor, rmotor                 : OUT      STD_LOGIC);
END motor_control;

ARCHITECTURE a OF motor_control IS
     SIGNAL count_motor: STD_LOGIC_VECTOR( 4 DOWNTO 0 );
BEGIN
     PROCESS
          BEGIN
                                        -- Count_motor is a 20ms  timer
               WAIT UNTIL clock_1Khz'EVENT AND clock_1Khz = '1';
                         IF count_motor /=19 THEN
                              count_motor <= count_motor + 1;
                         ELSE
                              count_motor <= "00000";
                         END IF;
                         IF count_motor >= 17 AND count_motor < 18 THEN
                                        -- Don't generate any pulse for speed = 0
                                        IF lmotor_speed = '0' THEN
                                             lmotor <= '0';
                                        ELSE
                                             lmotor <= '1';
                                        END IF;
                                        IF rmotor_speed = '0' THEN
                                             rmotor <= '0';
                                        ELSE
                                             rmotor <= '1';
                                        END IF;
                                        -- Generate a 1 or 2ms pulse for each motor
                                        -- depending on direction
                                        -- reverse directions between the two motors because
                                        -- of servo mounting on the UP1-bot base
                         ELSIF count_motor >=18 AND count_motor <19 THEN
                                        IF lmotor_speed /= '0' THEN
                                             CASE lmotor_dir IS
                                                  -- FORWARD
```

```
                                WHEN '0' =>
                                        lmotor <= '1';
                                -- REVERSE
                                WHEN '1' =>
                                        lmotor <= '0';
                                WHEN OTHERS => NULL;
                                END CASE;
                        ELSE
                                lmotor <= '0';
                        END IF;
                        IF rmotor_speed /= '0' THEN
                            CASE rmotor_dir IS
                                -- FORWARD
                                WHEN '1' =>
                                        rmotor <= '1';
                                -- REVERSE
                                WHEN '0' =>
                                        rmotor <= '0';
                                WHEN OTHERS => NULL;
                                END CASE;
                        ELSE
                                rmotor <= '0';
                        END IF;
                ELSE
                        lmotor <= '0';
                        rmotor <= '0';
                END IF;
    END PROCESS;
END a;
```

12.5 Sensors for the UP1-bot

A wide variety a sensors can be attached to the UP1-bot. A few of the more interesting sensors are described here. These include infrared modules to avoid objects, track lines, and support communication between UP1-bots. Other modules include sonar to measure the distance to objects, and a digital compass to determine the orientation of the UP1-bot.

Signal conditioning circuits are required in many cases to convert the signals to digital logic levels for interfacing to the digital inputs and outputs on the UP 1 board. Analog sensors will require an analog to digital converter IC to interface to the UP 1 board, so these devices pose a more challenging problem.

Sensor module kits are available and are the easiest to use since they come with a small printed circuit board to connect the parts. Sensors can also be built using component parts and assembled on a small protoboard attached to the UP1-bot. Sensor modules are interfaced by connecting jumper wires to digital inputs and outputs on the UP 1 board FLEX expansion B female header socket. Regulated +5V and unregulated +7.2V power is also available on the FLEX header socket.

- **Line Tracker Sensor**

A line tracker module from Lynxmotion is shown in Figure 12.3. This device uses three pairs of red LEDs and infrared (IR) phototransistor sensors that indicate the presence or absence of a black line below each sensor. When the correct voltages are applied in a circuit, an IR phototransistor operates as a switch. When IR is present the switch turns on and when no IR is present the switch turns off. The LED transmits red light that contains enough IR to trigger the phototransistor.

Each LED and phototransistor in a pair are placed at a 45-degree angle relative to the floor so that the light from the LED bounces off the floor and back to the IR phototransistor. The LED and IR sensor must be mounted very close to the floor for reliable operation. Black tape or a black marker is used to draw a line on the floor. The black line does not reflect light so no IR signal is returned.

Three pairs of LEDs and IR phototransistor sensors produce the three digital signals, left, center, and right. The UP1-bot can be programmed to follow a line on the floor by using these three signals to steer the robot. The mail delivery robots used in large office buildings use a similar technique.

Figure 12.3 Line Tracker Module – 3 LEDs and Phototransistors are mounted on bottom of board

- **Infrared Proximity Detector**

An IR proximity sensor module from Lynxmotion is seen in Figure 12.4. The UP1-bot can be outfitted with an infrared proximity detector that is activated by two off-angle infrared transmitting LEDs. The circuit utilizes a center-placed infrared sensor (Sharp GP1U5) to detect the infrared LED return as seen in Figure 12.5. The Sharp GP1U5 was originally designed to be used as the IR receiver in TV and VCR remote control units. From the diagram, one can see that the sensitivity of the sensor is based on the angle of the LEDs. The LED's can be outfitted with short heat-shrink tubes to better direct the infrared light forward. This prevents a significant number of false reflections coming from the floor. The IR sensor will still occasionally detect a few false returns and it will function more reliably with some hardware or software filtering.

Figure 12.4 IR Proximity Sensor Module – Two IR LEDs on sides and one IR sensor in middle.

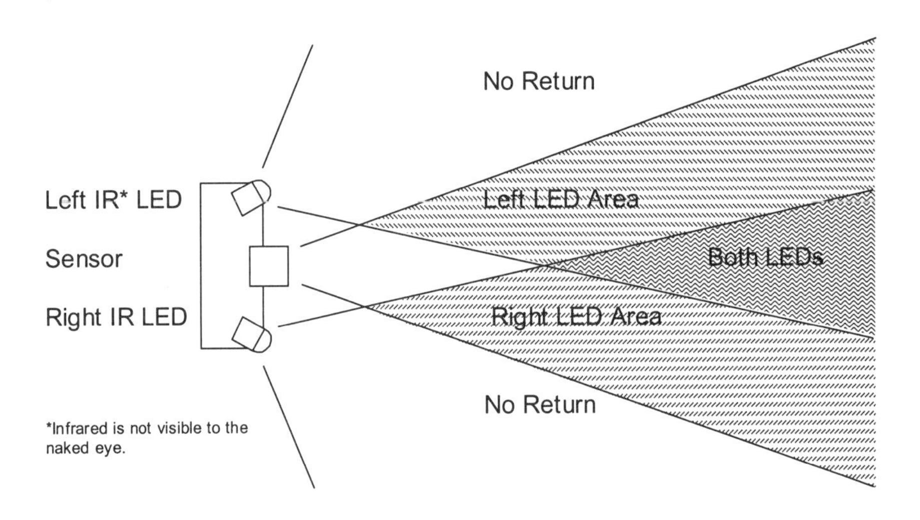

Figure 12.5 Proximity detector active sensor area.

As seen in Figure 12.6, the circuit on the IR proximity module utilizes a small feedback oscillator to set up a transmit frequency that can be easily detected by the detector module. This module utilizes a band-pass filter that essentially filters out ambient light.

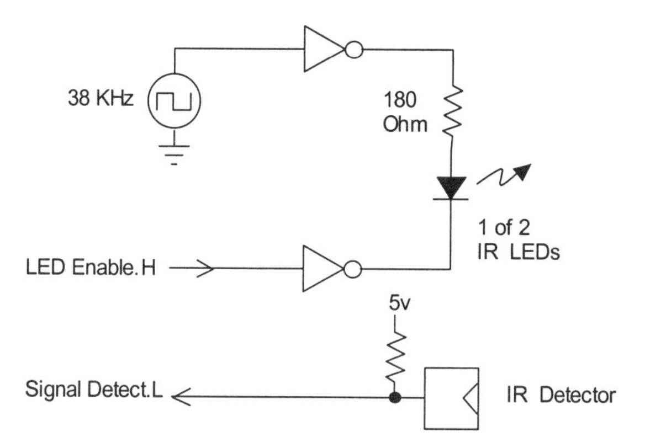

Figure 12.6 Circuit layout of one LED and the receiver module on the infrared detector.

In Figure 12.6, when the Left_LED Enable signal is High, the Low side of the IR LED is pulled to ground. This forces a voltage drop across the LED at the frequency of the 5v to ground oscillating signal. In other words, the LED produces IR light pulses at 38KHz. Using a 38Khz signal helps reduce noise from other ambient light sources.

Since the IR detector has an internal band-pass filter centered at 38KHz, the detector is most sensitive to the transmitted oscillating light. The 5v pull-up resistor allows the IR Detector's open collector output to pull up the SOUT signal to High when no IR output is sensed. To detect right and left differences, the right and left LEDs are alternately switched so that the detected signals are not ambiguous. If both the left and right LEDs detect an object at the same time, the object is in front of the sensor.

If the IR sensor was built from component parts, a hardware timer implemented on the UP 1 board could be used to supply the 38-40Khz signal. Similar IR LEDs and IR detector modules are available from Radio Shack, #276-137B, and Digikey, #160-1060. Assuming two UP1-bots are equipped with IR sensor modules, it is also possible to use this module as a serial communication link between the robots. One UP1-bot transmits using its IR LED and the other UP1-bot receives it using its IR sensor. To prevent interference, the IR LEDs are turned off on the UP1-bot acting as a receiver. Just like an IR TV remote, the IR LED and sensor must be facing each other. Bandwidth is limited by the 38Khz modulation on the IR signal and the filters inside the IR detector. (An IR sensor strip that converts IR to visible light is available from Radio Shack. This sensor can be used to confirm the operation of IR LEDs. If the Lynxmotion IR module is used, consider replacing the male header with a female machined pin strip socket. This makes it easier to connect jumper wires to the UP 1 board.)

- **Sonar Ranging Unit**

The Polaroid Sonar Module is shown in Figure 12.7. This device uses ultrasonic sound waves to measure distances from six inches to around 35 feet. These modules were originally developed to automatically focus Polaroid cameras. They are widely used in robotics. The output from the device is a digital echo signal. The timing of the echo indicates distance to the nearest object. The transducer first functions as a transmitter by emitting sixteen cycles of a 49Khz signal, and then functions as a receiver to detect sound waves returned by bouncing off nearby objects. Even though ultrasound is inaudible, the transducer also generates a slight audible click each time the device transmits.

The time it takes for the 49Khz-echo signal to return can be measured using a counter on the UP 1 board. This time is converted to distance since sound travels out and back at .9ms per foot. Only around 10 samples per second are possible with the device since it takes time to charge-up the portion of the circuit that produces the pulse and time to wait for echoes to return. The device operates off +5V DC and it generates short 400V pulses to fire the transducer. Current draw is almost 2 amps for the short period of time that the transducer fires, so a large decoupling capacitor of around 1000uf should be used between +5 and ground, or a separate 5VDC 7805 voltage regulator chip could be connected directly to the UP1-bot battery.

Other Sonar type sensors can be developed using low-cost electronic tape measuring units. Detailed information on these commercial devices is difficult to find. Many of the devices drive outputs only to Liquid Crystal Displays (LCDs), and these can be more difficult to interface to the UP 1 board.

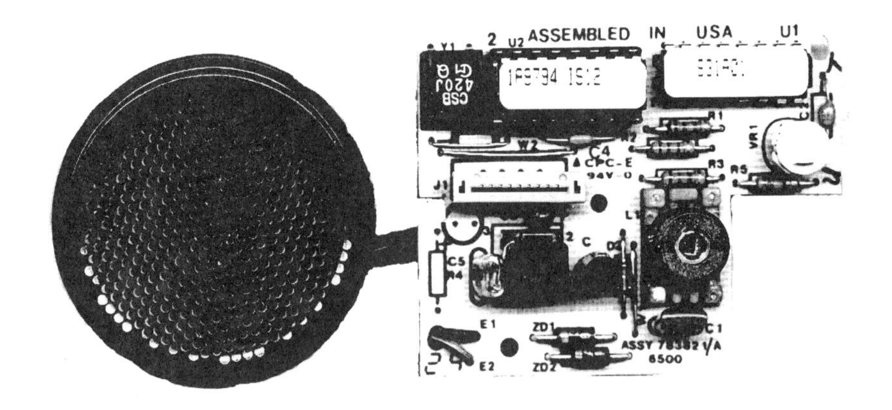

Figure 12.7 Polaroid Sonar Module – Left: Sonar Transducer and Right: Controller Board.

- **IR Distance Sensor**

The Sharp GPD2D02 seen in Figure 12.8 is an IR device that can provide distance measurements similar to the more expensive SONAR sensor. This sensor has a shorter range of 10 to 80cm (~ 4 to 32 inches). The distance is output by the sensor on a single pin as a digital 8-bit serial stream.

Figure 12.8 Sharp IR Ranging Module

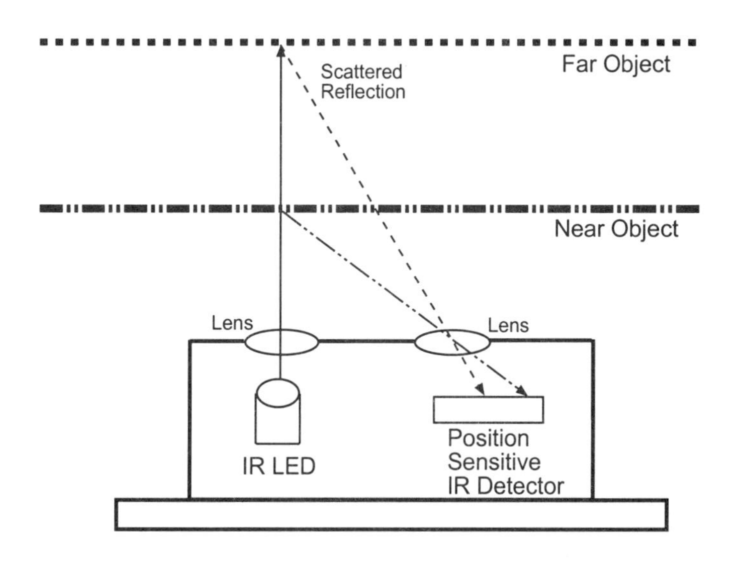

Figure 12.9 Operation of Sharp IR Ranging Module

As shown in Figure 12.9, internally the GPD2D02 contains an IR LED and a position-sensitive IR detector. The IR LED transmits a modulated beam of infrared light. When the light strikes an object, most of the light will be reflected back to the LED. Since no surface is a perfect optical reflector, scattering of the IR beam occurs at the surface of the object and some of the light is reflected back to the position sensitive detector. By comparing the near and far object beams shown in figure 12.9, it is apparent that position at which the scattered reflected IR beam hits the detector is a function of the reflection angle.

The 8-bit integer value reported by the sensor in cm is approximately

$$1000 * \tan^{-1}\left(\frac{1.9}{DISTANCE}\right) + offset\,.$$

The 1.9.cm term is the distance between the lenses. The offset is the no-object present value returned by the sensor. This offset constant can vary by as much as 17 between different sensors and has a typical value of 25. Note that a close object reports a larger value and a distant object reports a smaller value. Objects closer than 10cm will report an incorrect value and should be avoided by placing the sensor away from the edge of the robot. Large objects beyond 80 cm can sometimes report an incorrect value that makes them appear closer.

A special connector (Japan Solderless Terminal #S4B-ZR) is required to connect to the GPD2D02. If desoldering equipment is available, the small connector can also be desoldered from the sensor and wires attached directly to the sensor.

In addition to +5 and ground pins, the sensor has an input, Vin, and a serial output, Vout. Vin is an input to the sensor that clocks out the serial data on Vout. When Vin is low for around 70 milliseconds (ms), the sensor takes a reading. When a reading is available, Vout goes high.

On each of the successive next eight falling clock edges of Vin, the sensor will output a new data bit. The eight data bits should be clocked in to the CPLD on the rising edges of Vin when they are stable. When clocking out the data, the clock period on Vin should be 0.4 ms or less. The eight data bits are clocked out in high to low order. If Vin is not dropped low within 1.5 milliseconds after clocking out the final data bit, the sensor shuts down to save power. A shift register can be used to assemble the data bits. The demo program IR_DIST.GDF on the CD-ROM contains a VHDL based IP core for use with the GP2D02 sensor.

The sensor's Vin pin is an open drain input. Open drain or open collector inputs should never be driven high. An FPLD's tri-state output pin can be connected directly to an open drain input, if the tri-state output is never driven high. When Vin should be high, tri-state the FPLD's output pin and when the output should be low, drive the output pin low with the tri-state gate turned on with a low output.

Open drain or open collector inputs contain an internal pull-up resistor to +5. Multiple open drain (open collector) outputs can be tied together to a single open drain (open collector) input to perform a wired-AND operation. Any one of the outputs can pull the input low. If no output pulls the signal low, a single pull-up resistor forces the input high. This wired-AND operation occurs just by tying the open drain (open collector) outputs together and no physical AND gate is needed. In negative logic, a wired-OR operation occurs.

Normal gate outputs cannot be connected. This wired-AND logic only works because these gates have special output circuits do not contain a transistor that forces the input high. This transistor is present in normal gate outputs. If a normal gate output is connected to other open drain (open collector) outputs, it's transistor could turn on to force the input high at the same time another gates' output transistor turns on to force it low. This would short the power supply to ground drawing excessive current that might damage the devices.

- **Magnetic Compass Sensor**

Various electronic components are available that detect the magnetic field of the earth to indicate direction. A low-cost digital compass sensor is shown in Figure 12.10. The Dinsmore model 1490, often used in electronic automobile compasses, is a combination of a miniature rotor jewel suspended with four Hall-effect (magnetic) switches. Four active-low outputs are provided for the four compass directions. When the module is facing North, the North output is Low and the other three outputs will be High. Eight directions are detected by the device, since two outputs can become active simultaneously. In this way, the device can indicate the four intermediate directions, NE, SE, SW, and NW. NE for example activates the active-low North and East outputs. The device can operate off +5V.

Four 2.2K ohm pull-up resistors to +5V are required to interface to the UP 1 board since the four digital output pins, N, S, E, and W, all have open collector outputs. Just like a real compass, a time delay is needed after a quick rotation to allow the outputs to stabilize.

An analog version of the device is available with 1-degree accuracy, but it requires an analog-to-digital conversion chip or signal phase timing for interfacing. If the compass module leads are carefully bent, the compass module and the four required pull-up resistors can be mounted on a standard 20-pin DIP, machined-pin, wire-wrap socket and connected to the UP 1 header socket.

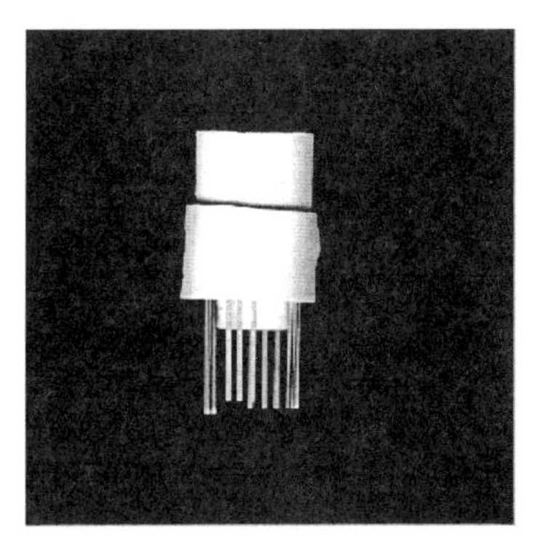

Figure 12.10 Dinsmore 1490 Digital Compass Sensor.

12.6 Assembly of the UP1-bot Body

Assembly of the UP1-bot can be accomplished in about an hour. A drill or drill press, screwdriver, scissors, a soldering iron, and a wire stripper are the only tools required. First, obtain the parts in the parts list. Next, drill out the holes in the round Plexiglas base (part #15) as shown in Figure 12.11. To prevent scratches, leave the paper covering on the Plexiglas until all of the holes are marked and drilled out. The front of the base is on the right side in Figure 12.11. The wheel slots are symmetric with respect to the center of the circle.

Proper alignment of the four screw mounting holes for the UP 1 board is critical. Unscrew the four standoffs from the bottom of the UP 1 board. Carefully place it towards the rear of the plastic base as shown in Figure 12.11, and mark the location of the screw holes using a pen or pencil.

Locate the cable and switch holes as shown in Figure 12.11. Exact positioning on these holes is not critical. If one is available, use an automatic center punch to help align the drill holes. The FLEX chip should face towards the front of the base. The extra space in front of the UP 1 board on the plastic base is used for sensor modules. Re-attach the standoffs to the UP 1 board and set it aside. After all holes are drilled, remove the paper covering the Plexiglas.

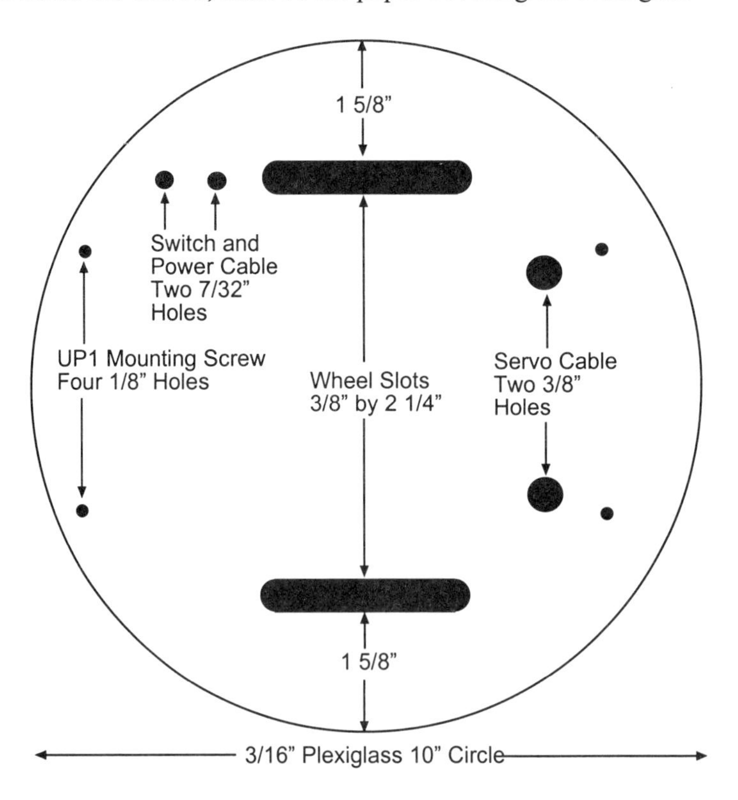

Figure 12.11 UP1-bot Plexiglas Base with wheel slots and drill hole locations.

Mount the toggle switch (part #8) in the hole provided in the base. If available, Loctite or CA glue can be used on the switch mounting threads to prevent the switch nut from working lose. Solder the red wire (+7.2V) from the battery connector to one of the switch contacts. This is the connector with wires that plugs into the battery pack connector (part #4). Solder one of the twin lead wires (part #9) to the other switch terminal. Solder the other twin lead wire to the black (GND) battery connector wire and insulate the splice with heat shrink tubing or electrical tape (part #10).

Route the twin lead wire through the hole provided in the base. A small knot in the twin lead on the bottom side of the base can be used for strain relief. Solder the power connector (part #11) to the other end of the twin lead wire on the top of the base. The center conductor is +7.2V and the outer conductor is ground on the power connector.

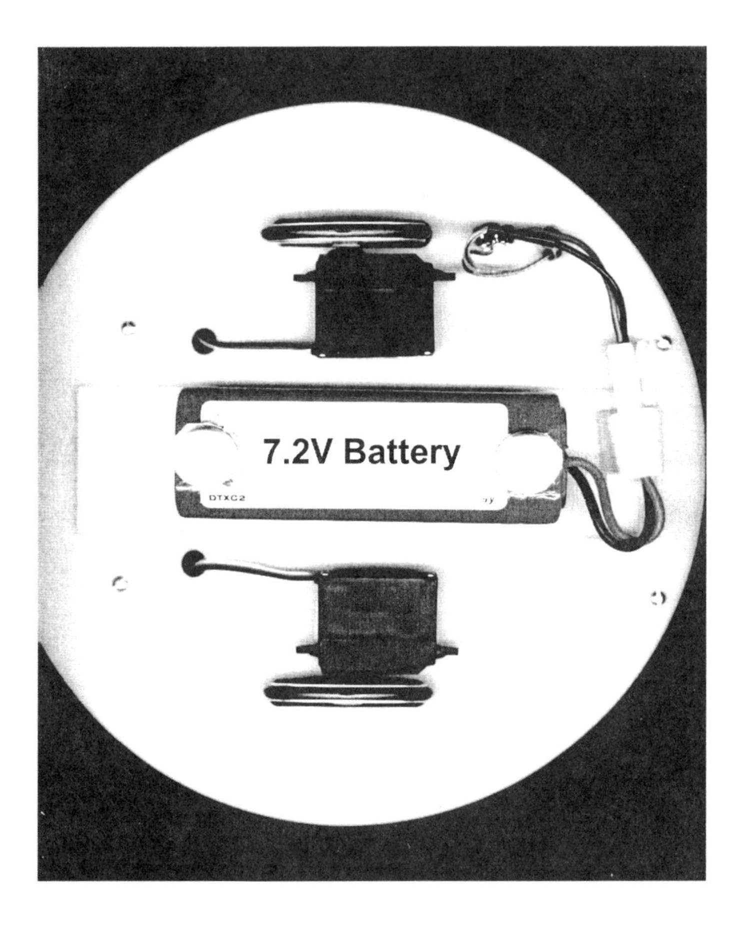

Figure 12.12 Bottom view of UP1-bot base showing battery, servos and cabling.

Check the power connections with an ohmmeter for shorts and proper polarity before connecting the battery. For strain relief and extra insulation, consider sealing up the power connector with Silicone RTV or insulating one of the wire connections with heat shrink tubing. Be careful, NiCAD batteries have been known to explode or catch on fire if there is a short. A fuse on the battery power wire might be a good idea, if you are prone to shorting out circuits.

Attach the battery pack (part #2) to the bottom of the Plexiglas base with sticky-back Velcro (part #16). Figure 12.12 is a close-up photo of the bottom side of the UP1-bot. The battery pack should mount in the middle of the base about one inch off center towards the rear, with the battery pack connector facing the rear.

The battery is moved towards the rear for balance to place the weight on the rear skid. The Velcro on the base should be around 2 inches longer than the battery pack towards the rear of the robot to allow for positioning of the battery later on to balance the robot. The wire and connector on the battery pack should also be attached to the base to prevent it from dragging on the floor. Attach a small piece of Velcro on the rear of the connector so that the battery wires can be attached to the base. Attach the battery pack to the base.

Solder a 60-pin female header socket (part #7) to the FLEX expansion B location. Attach the UP 1 board to the top of the base with 4-40 screws (part #18), using the hex spacers provided on the UP 1 board (part #14). Figure 12.13 is a close-up photo of the top of the UP1-bot. Double check power connections and polarity with an ohmmeter. The inner contact on the power connector should be +7.2V, the outer contact is ground, and the toggle switch should turn it off. Then plug the power connector into the UP 1 board. Plug in the battery connector and flip the power switch. A green LED should light up on the UP 1 board indicating power on. The FLEX expansion B header socket faces the front of the robot.

Mount the wheels (part #5) on two modified servos (part #3). Enlarge the hole in the center of each wheel by drilling it out partially with a drill bit that is the same size as the servo output shaft. The depth of the hole should be slightly shorter than the servo output shaft and not all the way through the wheel, so that the wheel does not contact the servo body. The servo output shaft screw is inserted on the side of the wheel with the smaller hole. A washer may be required on the servo screw. The wheel should not contact the servo case and must be mounted so that it is straight on the servo. CA glue or Blue Loctite can also be used to attach the wheels and screws more securely to the servo output shaft.

Attach the servos to the bottom of the base using foam tape (part #17). The servo body faces toward the center of the base. Be sure to carefully center the wheels in the plastic-base wheel slot. Make sure all surfaces are clean and free of grease, so that the foam tape adhesive will work properly. Lightly sanding the servo case and adding a drop of CA glue helps with tape adhesion. Route the servo connector and wire through the holes provided in the base.

Attach a skid (part #19) at the rear of the battery pack using layers of foam tape as needed. Move the battery as needed so that the robot rests on the two wheels and the rear skid.

Attach another skid to the front of the battery pack using several layers of foam tape. The front skid should not contact the floor and at least ¼ inch of clearance is recommended. The front skid only serves to prevent the robot from tipping forward during abrupt stops.

Attach the 2-pin header (part #12) to the red (+5V) and black (GND) wires on the servo connector. These can be plugged in each end of the FLEX expansion B header socket. Pins 3 (+5V) and 4 (GND) for the right servo and pins 57 (+5V) and 58 (GND) for the left servo. Some manufacturers' servos have different power connections, but they all have three pins. Wrap extra servo wire around the hex spacers underneath the UP 1 board. Insert two jumper wires into the remaining signal contact on each servo connector. This is typically a white or yellow wire. For the left servo, plug the other end of the jumper into pin 16 (FLEX Pin 110). For the right servo, plug the other end of the jumper wire into pin 56 (FLEX pin 162). Assembly of the UP1-bot body is now complete.

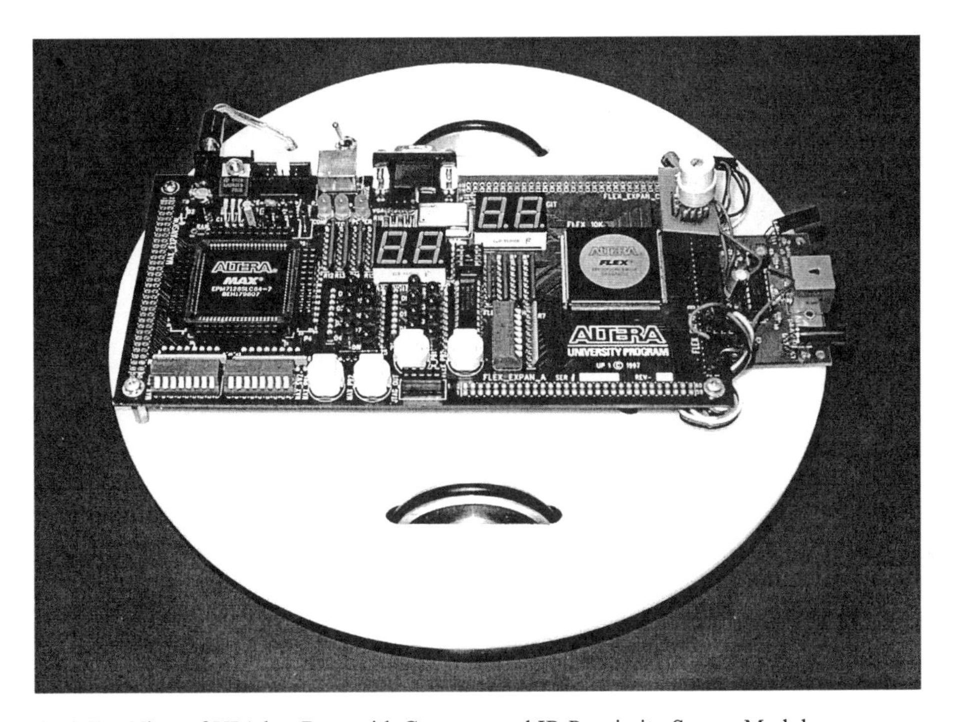

Figure 12.13 Top View of UP1-bot Base with Compass and IR Proximity Sensor Modules.

Optional sensor modules such as the IR proximity detector or line tracker can be attached to the base unit with foam tape. Run jumper wires from the sensors to the FLEX expansion B header socket.

- **Parts List for the UP1-bot**

1. An Altera UP 1 Board. The UP 1 board serves as the controller for the UP1-bot. It is attached to the UP1-bot body with screws. No modifications are required to the board other than adding a 60-pin female header socket to the FLEX expansion B location.

- **Parts Available from a Hobby Store**

2. **A 7.2V 1300-1700mAh Rechargeable NiCAD battery pack with the standard Kyosho battery connector.** This is a standard R/C car part, and it is used to power the UP1-bot. The battery will need to be charged prior to first use.

3. **Two modified R/C Servomotors**. Two identical model servos are required so that the motors run at the same speed. Servo modifications are described in section 12.3. Any servo should work. The following servos have been tested: Tower Hobbies TS53J, Futaba S148 and S3003, and HS 300. Some manufacturers' servos appear to run in the reverse direction. This is easily fixed in the hardware design since the motor controller is implemented on the UP 1 board.

4. **Kyosho Female Battery connector with wire leads, Duratrax or Tower Hobbies #DTXC2280**. This is used to connect to battery. A connector is needed so that the battery can be disconnected from the UP1-bot and connected to a charger.

5. **Two Prather Products 2¼-inch aluminum racing wheels with rubber O ring tires, Tower Hobbies #PRAQ1810**. These are attached to the servos and used as wheels. Hayes Products #114, 2 ¼-inch hard plastic racing wheels (also available from Tower Hobbies) can be used as a substitute.

6. **A charger for the 7.2V battery pack**. An adjustable DC power supply can be used to charge the battery if it is properly adjusted and timed so that the battery is not overcharged. Overcharged batteries will get hot and will have a shorter life. Automatic peak-detection quick chargers are the easiest and most foolproof to use. These chargers shut off automatically when the battery is charged. One quick charger can be used for several robots as a full charge is achieved in less than 30 minutes with around 5 Amps maximum charge current. Inexpensive trickle battery chargers deliver only around 75 mA of charge current, and they will require several hours charge the battery.

- **Parts Available from an Electronics Parts Store**

7. **A 60-pin .1-inch double row female header socket, DigiKey #929975-01-36 or equivalent**. The socket is soldered into FLEX expansion connector B on the UP 1 board. This is used to connect servos and sensors to the UP 1 board. Longer 72-pin header sockets can be cut off to 60 pins.

8. **A miniature toggle switch with solder lug connections.** The switch should have a contact rating of more than two amps (Radio Shack #275-635B or equivalent). Only two contacts or single pole single throw (SPST) is needed on the switch to turn power on and off.

9. **Approximately 6 inches of small-gauge twin-lead speaker wire.** This part is used to connect power to the UP 1 board. The wire must fit into the DC power plug (part# 11). Typically, 20-22 gauge wire is required. Two individual wires can also be used, but twin lead is preferred.

10. **A 1-inch piece of small heat shrink tubing or electrical tape.** This part is used to insulate a splice in the twin-lead power wire.

11. **A Coaxial DC Power Plug with 5.5mm O.D. and 2.5mm I.D., Radio Shack Number 274-1568C** or equivalent. This power plug fits the power socket on the UP 1 board.

12. **Two 2-pin .1-inch Male Headers**. This part is used to connect servo power leads to the UP 1 FLEX expansion B female header. Wire jumpers can also be used, but this is more reliable. These can be cut off a longer header.

13. **An assortment of small wire jumpers**. These are the jumper wires commonly used for protoboards. Two short jumpers are used to connect the two servo signal wires, and other jumpers are used to any connect sensor boards.

14. **Four, 1-inch hex spacers with 4-40 threads.** These are needed only if the UP 1 board being used does not have mounting spacers with 4-40 screw treads in the bottom. A few of the UP 1 boards have threaded rubber mounts instead of hex spacers with threads on both ends.

- **Parts Available from a Hardware Store**

15. **3/16-inch thick Plexiglas cut into a 10-inch diameter circle**. This part is the base of the robot. Colored Plexiglas such as opaque white, will not show scratches as easy as clear. Holes to cutout and drill are shown in Figure 12.11. If a band saw, jig saw, or other machine tool is not available, a local plastics fabricator can cut this out. When using a number of very large sensors, it may be necessary to increase the size slightly or add another circular deck for sensor mounting. A larger robot requires more space for maneuvering. To prevent scratches on the Plexiglas, keep the paper backing on the plastic until all of the holes have been marked and drilled out.

16. **One 8-inch long strip of 2-inch wide sticky-back Velcro**. Two 8-inch long strips, 1 inch wide can also be used. The Velcro is used to attach the battery to the bottom of the Plexiglas base. Since the battery is attached with Velcro and a connector, it can be quickly replaced and removed for charging.

17. **Approximately 8 inches of 1-inch wide double-sided 3M foam tape**. This is used to attach servos, skids, and optional sensor boards to the base. Be sure to clean surfaces to remove any grease or oil prior to application of the tape for better adhesion.

18. **Four 4-40 Screws 5/16-inch or slightly longer**. The screws are used to attach the UP 1 board to Plexiglas. The screws thread into the hex spacers attached to the UP 1 board.

19. **Two small Teflon or Nylon Furniture Slides**. Magic Sliders 7/8-inch diameter circular discs work well. The slides are used as a skid instead of a third wheel on the UP1-bot. Metal or hard plastic will also work. Attached to the bottom of the battery with several layers of foam tape, the optional front skid is used for stability during abrupt stops. A Teflon skid actually works better than a small caster.

20. **Blue Loctite, Cyanoacrylate (CA) Glue, and Clear Silicone RTV.** These adhesives and glues are useful to secure screws, servos, and wheels. The mechanical vibration on moving robots tends to shake parts loose over time. These items can also be found at most hobby shops. Only a few drops are needed for a single robot. A single tube or container will build several robots.

12.7 UP1-bot FLEX Expansion B Header Pins

Motors and sensors are attached to the FLEX expansion B header. FLEX device pin numbers can be found in Table 12.1. Power connections are also available at each end of the header. Pin 1 is the unregulated power to the UP 1 board. In the UP1-bot, this will be the 7.2V output of the battery pack. If a sensor module has it's own on-board +5V regulator, use pin 1 for power. Suggestions for UP1-bot servo motor and sensor connections are shown in parenthesis.

Table 12.1 UP1-bot FLEX Expansion B Header Pins

Header Pin	FLEX Pin	Header Pin	FLEX Pin
1	Unregulated Power	2	GND
3	VCC	4	GND
5	VCC (Right Servo VCC)	6	GND (Right Servo VCC)
7	No connection	8	DI1/99
9	DI2/92	10	DI3/210
11	DI4/212	12	DEV_CLR/209
13	DEV_OE/213	14	DEV_CLK2/211
15	109	16	110 (Right Servo Signal)
17	111	18	113
19	114	20	115
21	116	22	117
23	118	24	119
25	120	26	126
27	127	28	128
29	129	30	131
31	132	32	133
33	134	34	136
35	137	36	138
37	139	38	141
39	142	40	143
41	144	42	146
43	147	44	148
45	149	46	151
47	152	48	153(IR Sensor Out)
49	154(IR Right LED)	50	156(Compass West)
51	157(IR Left LED)	52	158(Compass East)
53	159	54	161(Compass South)
55	162(Left Servo Signal)	56	163(Compass North)
57	VCC (Left Servo VCC)	58	GND (Left Servo GND)
59	VCC	60	GND
Note: Pins in parenthesis are used for UP1-bot Interfacing			

12.8 An Alternative UP 1 Robot Project Based on an R/C Car

A second option for building a UP 1 driven robot involves modifying a low-cost radio-controlled (R/C) car or truck. Fundamentally, almost any large R/C car or truck can be modified to work with the Altera board, although some are clearly better choices than others. In our robot, we used a Radio Shack (www.radioshack.com) R/C SUV shown in Figure 12.14. The R/C platform affords a more robust drive train and control; however, turning radius and noise levels are sacrificed over the smaller UP1bot. Following are some R/C car selection considerations that will affect available modifications and control of the new platform.

Figure 12.14 UP 1 Controlled R/C Car with IR Distance Sensors.

- ### Seven-Function Controls

When choosing an R/C car, select one that has a remote control with at least seven remote functions (forward, backward, forward-right, forward-left, backward-right, backward-left, and stop). Note that these low-cost R/C cars do not have variable speed or variable turning controls; however, once they are interfaced to the UP 1 board, variable speed and turning can be accomplished by changing the duty cycle of the command signals. (More on this later.)

A control module built using the UP 1's logic allows a relatively inexpensive R/C car to perform with the capabilities of the more expensive cars with "digital proportional steering" and "digital proportional speed controls." Once interfaced to the UP 1 board, an IP core (Robot_CTL) is used to handle control of all direction and speed control outputs. As illustrated in Figure 12.15, the IP core control module affords a higher degree of control than the original radio control. The outputs connect to the R/C cars internal control circuits that drive the DC Motors.

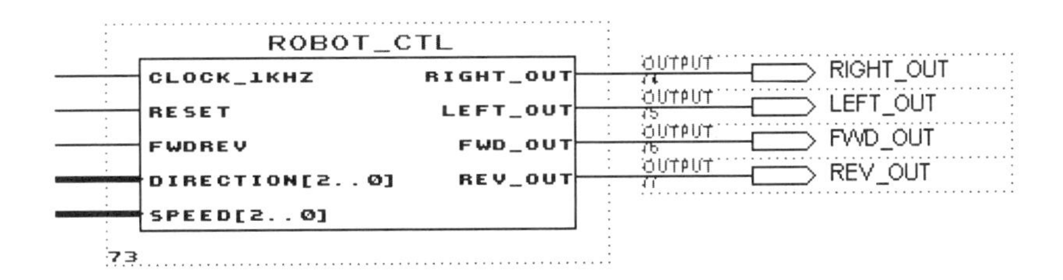

FwdRev	1 Bit	0 = Forward/1 = Reverse
Direction	3 Bits	First bit Left/Right, 2nd and 3rd bit is angle.
		0-00 = Left – Straight*
		0-01 = Left – Slight Turn
		0-10 = Left – Medium Turn
		0-11 = Left – Full Turn
		1-00 = Right – Straight*
		1-01 = Right – Slight Turn
		1-10 = Right – Medium Turn
		1-11 = Right – Full Turn
		* Note: 000 and 100 are both Straight
Speed	3 Bits	000 = Stop
		001 = Slowest Speed
		: : :
		111 = Fastest Speed

Figure 12.15 Robot Control IP Core with Pulsed Speed & Steering Control.

- **Speed**

When considering the speed of the vehicle, a modest speed is more desirable than the faster speeds. At 800 feet per minute, our prototype UP 1 controlled robot car moves fast enough to be difficult to catch. In almost all cases, the robot is operated at half the maximum speed or less. The limiting factor is generally the delay inherent in the sensor's input sampling rate and range. A fast moving car typically will hit the wall before a collision sensor can take the data samples needed to initiate avoidance.

To control the speed (or the degree of turn) a repeated pulse train is sent to the forward or reverse signals (left or right signals for direction). Instead of a steady high signal causing the car to move forward at full speed, the pulse train varies the duty cycle to change speed (or degree of turn). By modulating the duty cycle of the pulse train, "digital proportional control" can be implemented on each control signal. In other words, changing the duty cycle can control the speed and the degree of turn. The more the duty cycle approaches 100%, the harder the turn and/or the faster the speed.

The frequency of the pulsed control signal used must be higher than the natural mechanical frequency response of the system. A very slow changing pulse will cause the motor and gears to vibrate and make additional noise. Pulse frequencies of a few KHz are typically used to avoid this problem.

When reversing direction on a moving DC motor, it is common practice to include a small time delay with the motor turned off to reduce the inductive voltage spikes produced by the motor windings. Recall that changing current flow through an inductor produces voltage. Without the delay in some circuits, these high voltage spikes can damage or reduce the life of the transistors controlling the motor. This delay can be incorporated in the IP control core, if needed.

Figure 12.16 illustrates the relationship between turn angle, speed, and duty cycle on the control signals. The figure implies that the duty cycle is linear, i.e., a 50% duty cycle produces half speed. Actually, the duty cycle is very non-linear and highly dependent on the type of car, size of the DC motors, and power. (An R/C car with a dying battery performs as if the duty cycle is considerably less.) By experimenting with patterns of the 16-bit speed and direction vectors used in the IP core controller, a more linear relationship can be established between the command bit patterns and the actual performance of the vehicle.

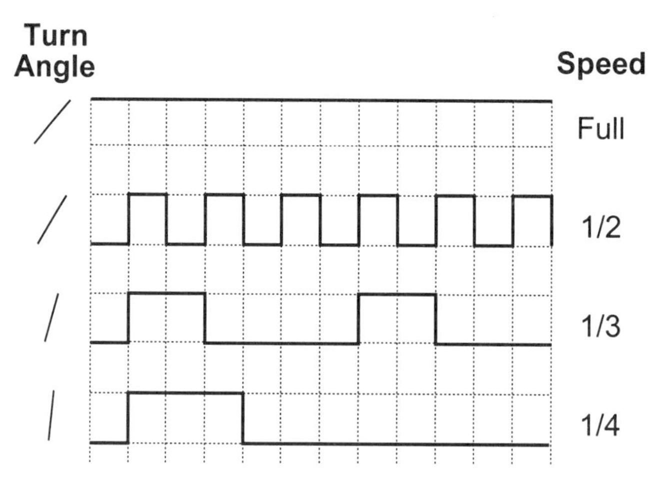

Figure 12.16 Affect of Duty Cycle on Turning Angle and Speed.

- **Battery Choice**

The choice of car will also dictate the type of batteries and charger that will be needed. Note that some cars come with 9.6V packs and others come with 7.2V packs. Both should work well with the UP 1 board as a controller. The prototype used the 7.2V pack that discharged quickly and required a second pack placed in parallel with the first to support longer run times. The cars with a 9.6V pack should give the UP 1 board's 5V onboard regulator a better regulator margin and a longer life between recharges.

- **Mounting the UP 1 Board**

  Before you select an R/C car, make sure that there is a good place to mount the UP 1 board. If the car is large enough, there is usually a large flat area under the car body cover molding to secure the UP 1 board.

- **Interfacing the UP 1 Board to the R/C Car**

  Remove appropriate body cover screws and expose the PC board receiver and control module. Most current R/C cars have a single electronic PC board that contains both control circuits. Generally, there is one 16 or 18-pin DIP radio command demodulator chip in the center of the board that converts the radio signals into simple digital control signals. These digital control signals then activate the H-bridge circuit that controls the DC motors that drive the wheels of the R/C car.

  An H-bridge is a standard electronic circuit used to control DC motors. It allows for both forward and reverse operation of the same DC motor. H-bridge circuits contain four large power transistors that are needed to turn on and reverse a DC motor.

> NEVER ATTEMPT TO DRIVE A MOTOR OR RELAY WITH AN FPLD PIN DIRECTLY, IT CANNOT SUPPLY THE HIGH CURRENT LEVELS NEEDED. THE FPLD'S OUTPUT PIN DRIVER CIRCUIT WOULD BE DESTROYED.

If the car supports seven functions, it will have at least four pins coming off of the DIP chip package that break down into Left, Right, Forward, and Reverse. Using a voltmeter or an oscilloscope, test which pins change when the remote control is set to each of the four directions. From each of the designated command pins on the chip, the trace on the PC board will run to separate H-bridge circuits for each motor.

By clipping or desoldering and pulling out the four control pins on the chip going to the board and soldering wires from each chip pin hole pad trace to the UP 1 (Figures 12.17 and 12.18), the UP 1 board can control the four directions and speed of the R/C car using the car's existing H-bridge circuits. In our modification, we desoldered the entire chip and put a socket on the board. To have the original control signals coming from the radio control module also sent to the UP 1 board, run four more wires from the clipped chip pins to the UP 1 header. If the four clipped chip pins (RC*x*) are connected to the UP 1 board, logic can be designed to include the original radio control functions supplied by the handheld remote control unit.

For those with a larger budget, higher quality R/C hobbyist cars are available with built-in proportional steering and pulsed electronic speed controls. The control interface to these cars uses standard R/C PWM signals that are identical to the PWM servo control bit described at the end of Section 12.2.

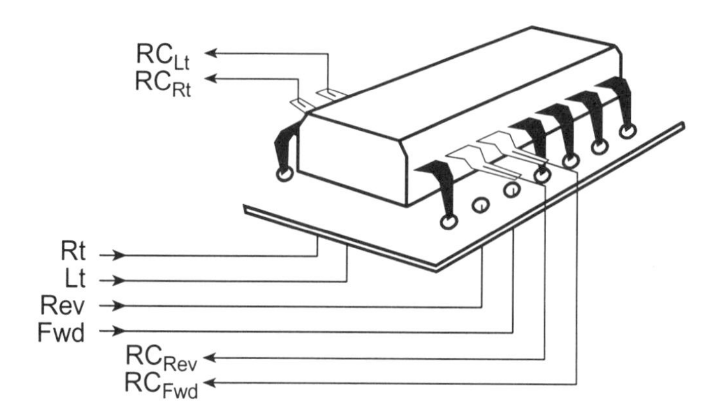

Figure 12.17 Interfacing to the R/C Car's Internal Control Signals at the Demodulator IC.

Figure 12.18 Photo Showing Control Modifications to R/C Car Control Board.

- **UP 1 Header Assembly for R/C Car**

  From the book's website, a UP 1 R/C car header printed circuit board (PCB) layout can be downloaded. This small board helps to facilitate robot power connections and sensor hookup. This plug-in UP 1 header module connects pins to a standard connector, which in turn simplifies the wiring to the R/C car and other sensors. Although not necessary, the header assembly also has its own regulated power and connections for a compass module, IR sensors, and an IR remote control.

12.9 For Additional Information

Radio-controlled cars and parts such as batteries, battery chargers, and servos can be obtained at a local hobby shop or via mail order from:

Tower Hobbies http://www.towerhobbies.com
P.O. Box 9078
Champaign, IL 61826-9078
800-637-6050

IR Proximity, Line Tracker, and Sonar sensor kits for robots can be obtained via mail order from:

Lynxmotion, Inc. http://www.lynxmotion.com
104 Partridge Road
Pekin, IL 61554-1403
309-382-1254

Mondotronics http://www.robotstore.com
4286 Redwood Highway #226
San Raphael, CA 94903
800–374–5764

IR sensors, data sheets, and Polaroid Sonar sensor kits can be obtained via mail order from:

Wirz Electronics http://wirz.com
P.O. Box 457
Littleton, MA 01460-0457
888-289-9479

The Sharp GP2D02 IR ranging sensor and cable, can be obtained from:

http://www.hobbyrobot.com/sensors/GP2D02.html

Polaroid Sonar sensor kits can also be obtained via mail order from:

Acroname http://www.acroname.com
P.O. Box 1894
Nederland, CO 80466
303-258-3161

Low-cost digital and analog compass sensors are available via mail order from:

Dinsmore Instrument Co. http://www.dinsmoregroup.com/dico
P.O. Box 345
Flint, Michigan 48501
810-744-1790

12.10 Laboratory Exercises

1. Develop a counter design to find the dead zone of each servo motor. The dead or null zone is the time near 1.5ms that actually makes the servo motor stop moving. As in the example motor driver code, send a width adjusted pulse every 20ms. You will need a resolution of at least .01ms to find the dead zone, so a clock faster than the example code is required. For example, the motor might actually stop at 1.54ms instead of 1.50ms. Use the clk_div UP1core function to provide the clock. The design should increase the width of the timing pulse if one pushbutton is hit and decrease the width if the other pushbutton is hit. Display the width of the timing pulse in the seven-segment LEDs. Use a FLEX DIP-switch input to select the motor to examine. By hitting the pushbuttons, you should be able to stop and reverse the motor. The dead zone will be between the settings where the drive wheel reverses direction. At the dead zone, the drive wheel should stop. Settings near the dead zone will make the motor run slower. Record the dead zone for both the left and right motor.

2. Using the dead zone settings from problem 1, design a motor speed controller. Settings within around .2ms of the dead zone will make the motor run slower. The closer to the dead zone the slower the motor will run. Include at least four speed settings for each motor. See if you can get the robot to move in a straight line at a slow speed.

3. Develop a speed controller for the robot drive motors by pulsing the drive motors on and off. The motors are sent a pulse of 1ms for reverse and 2ms for forward at full speed. If no pulse is sent for 20ms, the motor stops. If a motor is sent a 1 or 2ms pulse followed by no pulse in a repeating pattern, it will move slower. To move even slower use pulse, no pulse, no pulse in a repeating pattern. To move faster use pulse, no pulse, pulse in a repeating pattern. Using this approach, develop a speed controller for the robot with at least five speeds and direction. Send no pulse for the stop speed. Some additional mechanical noise will result from pulsing the motors at slow speeds. See if the robot will move in a straight line at a slow speed.

4. Use an IR LED and IR sensor to add position feedback to the motors. Some sensor modules are available that have both the IR LED and IR sensor mounted in a single plastic case. For reflective sensors, mark the wheels with radial black paint stripes or black drafting tape and count the pulses from the IR sensor to determine movement of the wheel. Another option would be to draw the radial stripes using a PC drawing program and print it on clear adhesive labels made for laser printers. The labels could then be placed on the flat side of the wheel. If a transmissive sensor arrangement is used, holes can be drilled in the main wheel or a second smaller slotted wheel could be attached to the servo output shaft that periodically interrupts the IR light beam from the LED to the sensor. In this case, the LED and sensor are mounted on opposite sides of the wheel. This same optical sensing technique is used in many mice to detect movement of the mouse ball. Use the position feedback to implement more accurate variable speed and position control for the motors.

5. Design a state machine using a counter/timer that will move the robot in the following fixed pattern:

   - Move forward for 6 seconds.
   - Turn right and go forward for 4 seconds (do not count the time it takes to turn).
   - Turn left and go forward for 2 seconds.
   - Stop, pause for 2 seconds, turn 180 degrees, and start over.

Determine the amount of time required for 90- and 180-degree turns by trial and error. A 10Hz or 100Hz clock should be used for the timer. Use the clk_div UP1core to divide the UP 1 on-board clock. The state machine should check the timer to see if the correct amount of time has elapsed before moving to the next state in the path. The timer is reset when moving to a new portion of the path. Use an initial state that turns off the motors until a pushbutton is hit, so that it is easier to control the robot during download. Since there is no motor position feedback, all turns and the actual distance traveled by the UP1-bot will vary slightly.

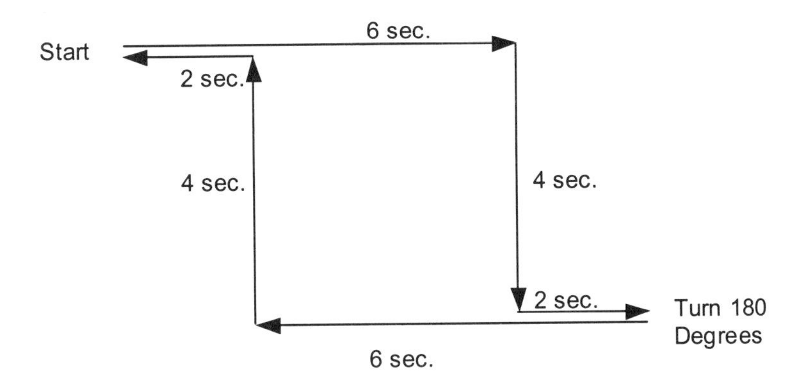

Figure 12.19 Simple path for state machine without sensor response.

6. Using a ROM with the LPM_ROM function, develop a ROM-based state machine that reads a motor direction and time from the ROM. Put a complex pattern such as a dance step in the ROM using a MIF file. For looping, another field in the ROM can be used to specify a jump to a different next address.
7. Using the keyboard UP1core, design an interface to the keyboard that allows the keyboard to be used as a remote control device to move the robot. Pick at least five different keys to command to robot to move, turn left, turn right, or stop.
8. Interface an IR proximity sensor module to the UP1-bot using jumpers connected to the FLEX female header socket. Attach the module in front of the header socket using foam tape. Use the raw power connection for the IR module's 9V input pin. Alternate driving the left and right IR LEDs at 100Hz. Check for an IR sensor return and develop two signals, LEFT and RIGHT to indicate if the IR sensor return is from the left or right IR LED. The IR LEDs may need to be adjusted or shielded with some heat shrink tubing so that the floor does not reflect IR to the sensor. Use the LEFT and RIGHT signals to drive the decimal points on the two seven-segment LEDs to help adjust the sensor. It may be necessary to filter the IR returns using a counter with a return/no return threshold for reliable operation. Using a clock faster than 100Hz, for example 10Khz, only set LEFT or RIGHT if the return was present for several clock cycles.
9. Using IR sensor input, develop a design for the UP1-bot that follows a person. The person must be within a foot or so of the UP1-bot. When a left signal is present turn

left, when a right signal is present turn right, and when both signals are present, move forward a few inches and stop. When all signals are lost, the UP1-bot should rotate until an IR return is acquired.

10. Use motor speed control and a state machine with a timer to perform a small figure eight with the UP1-bot.

11. Once the IR proximity sensor module from problem 8 is interfaced, design a state machine for the robot that moves forward and avoids obstacles. If it sees an obstacle to the left, turn right, and if there is an obstacle to the right, turn left. If both left and right obstacles are present, the robot should go backwards by reversing both motors.

12. With two UP1-bots facing each other, develop a serial communications protocol using the IR LEDs and sensors. Assume the serial data is fixed in length and always starts with a known pattern at a fixed clock rate. The IR LEDs are pulsed at around 40Khz and the sensor has a 40Khz filter, so this will limit the bandwidth to a few Khz. Transmit the 8-bit value from the FLEX DIP switches and display the value in the receiving UP1-bot seven-segment LED displays. Display the raw IR sensor input in the decimal point LED to aid in debugging and alignment.

13. Interface the line-following module to the UP1-bot, and design a state machine that follows a line. The line-following module has three sensor signals, left, center, and right. If the line drifts to the left, turn right, and if the line drifts to the right, turn left. Adjust turn constants so that the UP1-bot moves along the line as fast as possible. If speed control was developed for the UP1-bot as suggested in earlier problems, try using speed control for smaller less abrupt turns.

14. Using a standard IR remote control unit from a television or VCR and an IR sensor interfaced to the UP1-bot, implement a remote control for the UP1-bot. Different buttons on the remote control unit generate a different sequence of timing pulses. A digital oscilloscope or logic analyzer can be used to examine the timing pulses.

15. Interface the magnetic compass module to the UP1-bot, and design a state machine that performs the following operation:

    - Turn North.
    - Move forward 4 seconds.
    - Turn East.
    - Move forward 4 seconds.
    - Turn Southwest.
    - Move forward 6.6 seconds.
    - Stop and repeat when the pushbutton is hit.

    The compass has a small time delay due to the inertia of the magnetized rotor. Just like a real compass, it will swing back and forth for awhile before stopping. With care, the leads on the compass module can be plugged into a DIP socket with wire wrapped power supply and pull-up resistor connections on a small protoboard or make a printed circuit board for the compass with jumper wires to plug into the FLEX female header socket. Make sure the compass module is mounted so that it is level.

16. Interface a Sonar-ranging module to the UP1-bot and perform the following operation:
    - Scan the immediate area 360 degrees by rotating the robot and locate the nearest object.
    - Move close to the object and stop.
17. Attach the Sonar transducer to an unmodified servo's output shaft. Use the new servo to scan the area and locate the closest object. To sweep the unmodified servo back and forth, a timing pulse that slowly increases from 1ms to 2ms and back to 1ms is required. Move close to the nearest object and stop.
18. Attach several IR ranging sensors to the UP1-bot and use the sensor data to develop a wall following robot.
19. Interface additional sensors, switches, etc., to the UP1-bot so that it can navigate a maze. If several robots are being developed, consider a contest such as best time through the maze or best time after learning the maze.
20. Use the μP 1 computer from Chapter 8 to implement a microcontroller to control the robot instead of a custom state machine. Write a μP 1 assembly language program to solve one of the previous problems. Interface a time-delay timer, the sensors, and the motor speed control unit to the μP 1 computer using I/O ports as suggested in problem 8.6. The additional machine instructions suggested in the exercises in Chapter 8 would also be useful.
21. Develop and hold a UP1-bot design contest. Information on previous and current robotics contests can be found online at various web sites. Here are some ideas that have been used for other robot design contests:
    - Robot Maze Solving
    - Robot Dance Contest
    - Sumo Wrestling
    - Robot Soccer Teams
    - Robot Laser Tag
    - Fire Fighting Robots
    - Robots that collect objects

CHAPTER 13

A RISC Design: Synthesis of the MIPS Processor Core

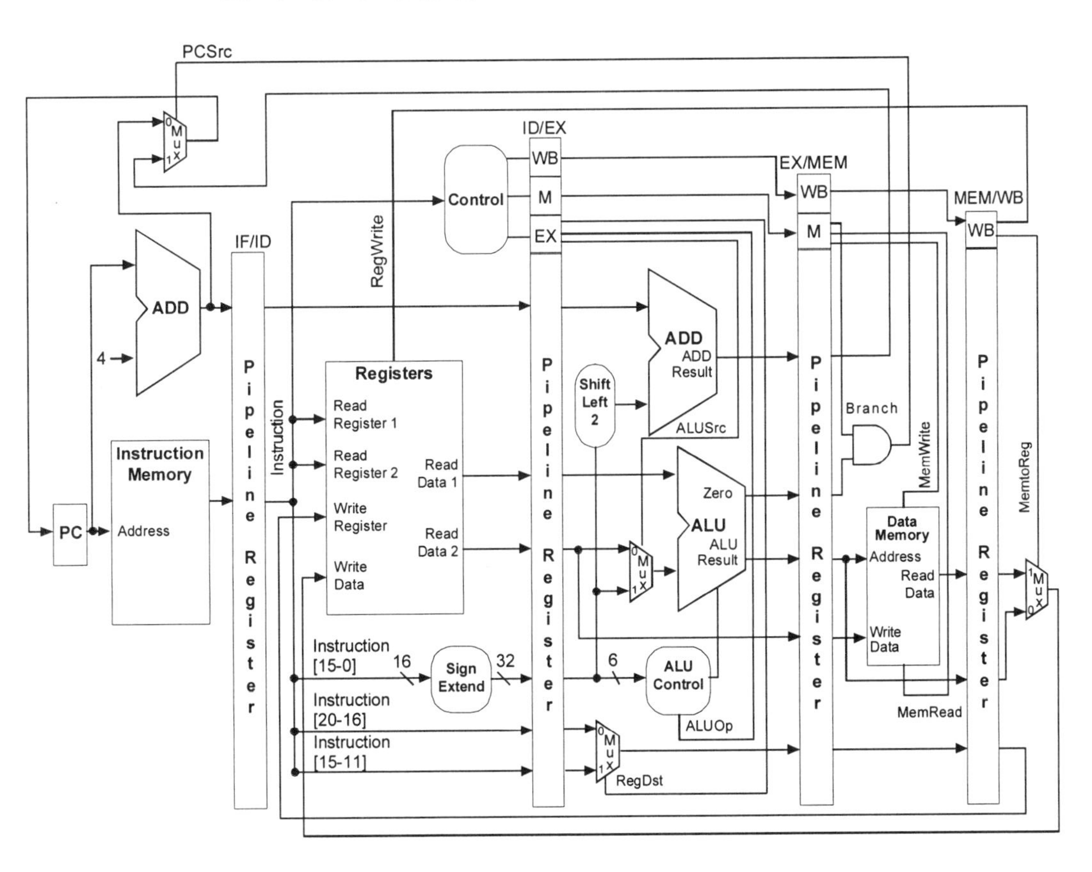

Block Diagram of the MIPS Processor Core

13 A RISC Design: Synthesis of the MIPS Processor Core

13.1 The MIPS Instruction Set and Processor

The MIPS is an example of a modern reduced instruction set computer (RISC) developed in the 1980s. The MIPS instruction set is used by NEC, Nintendo, Silicon Graphics, Sony, and several other computer manufacturers. It has fixed-length 32-bit instructions and thirty-two 32-bit general-purpose registers. Register 0 always contains the value 0. A memory word is 32 bits wide.

As seen in Table 13.1, the MIPS has only three instruction formats. Only I-format LOAD and STORE instructions reference memory operands. R-format instructions such as ADD, AND, and OR perform operations only on data in the registers. They require two register operands, Rs and Rt. The result of the operation is stored in a third register, Rd. R-format shift and function fields are used as an extended opcode field. J-format instructions include the jump instructions.

Table 13.1 MIPS 32-bit Instruction Formats.

Field Size	6-bits	5-bits	5-bits	5-bits	5-bits	6-bits
R- Format	Opcode	Rs	Rt	Rd	Shift	Function
I - Format	Opcode	Rs	Rt	Address/immediate value		
J - Format	Opcode	Branch target address				

LW is the mnemonic for the Load Word instruction and SW is the mnemonic for Store Word. The following MIPS assembly language program computes A = B + C.

```
LW $2, B            ;Register 2 = value of memory at address B
LW $3, C            ;Register 3 = value of memory at address C
ADD $4, $2, $3      ;Register 4 = B + C
SW $4, A            ;Value of memory at address A = Register 4
```

The MIPS I-format instruction, BEQ, branches if two registers have the same value. As an example, the instruction BEQ $1, $2, LABEL jumps to LABEL if register 1 equals register 2. A branch instruction's address field contains the offset from the current address. The PC must be added to the address field to compute the branch address. This is called PC-relative addressing.

LW and SW instructions contain an offset and a base register that are used for array addressing. As an example, LW $1, 100($2) adds an offset of 100 to the contents of register 2 and uses the sum as the memory address to read data from. The value from memory is then loaded into register 1. Using register 0, which always contains a 0, as the base register disables this addressing feature.

Table 13.2 MIPS Processor Core Instructions.

Mnemonic	Format	Opcode Field	Function Field	Instruction
Add	R	0	32	Add
Addi	I	8	-	Add Immediate
Addu	R	0	33	Add Unsigned
Sub	R	0	34	Subtract
Subu	R	0	35	Subtract Unsigned
And	R	0	36	Bitwise And
Or	R	0	37	Bitwise OR
Sll	R	0	0	Shift Left Logical
Srl	R	0	2	Shift Right Logical
Slt	R	0	42	Set if Less Than
Lui	I	15	-	Load Upper Immediate
Lw	I	35	-	Load Word
Sw	I	43	-	Store Word
Beq	I	4	-	Branch on Equal
Bne	I	5	-	Branch on Not Equal
J	J	2	-	Jump
Jal	J	3	-	Jump and Link (used for Call)
Jr	R	0	8	Jump Register (used for Return)

A summary of the basic MIPS instructions is shown in Table 13.2. In depth explanations of all MIPS instructions and assembly language programming examples can be found in the references listed in section 13.11.

A hardware implementation of the MIPS processor core based on the example in the widely used textbook, *Computer Organization and Design The Hardware/Software Interface* by Patterson and Hennessy, is shown in Figure 13.1. This implementation of the MIPS performs fetch, decode, and execute in one clock cycle. Starting at the left in Figure 13.1, the program counter (PC) is used to fetch the next address in instruction memory. Since memory is byte addressable, four is added to address the next 32-bit (or 4-byte) word in memory. At the same time as the instruction fetch, the adder above instruction memory is used to add four to the PC to generate the next address. The output of instruction memory is the next 32-bit instruction.

The instruction's opcode is then sent to the control unit and the function code is sent to the ALU control unit. The instruction's register address fields are used to address the two-port register file. The two-port register file can perform two independent reads and one write in one clock cycle. This implements the decode operation.

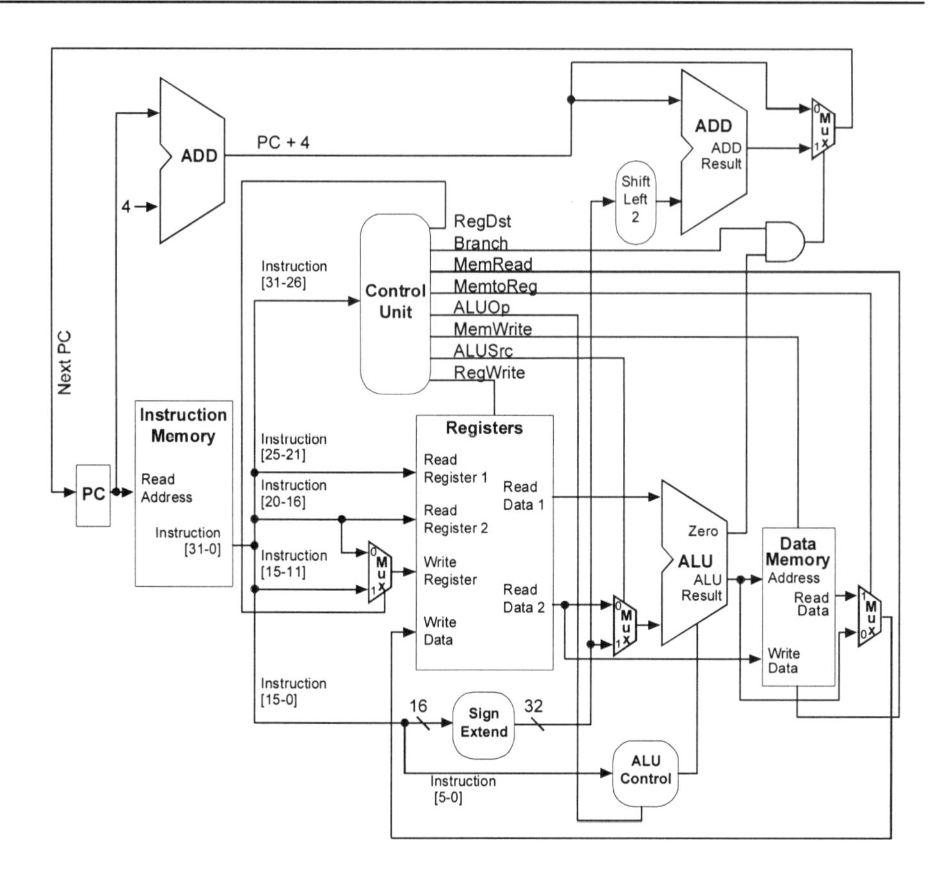

Figure 13.1 MIPS Single Clock Cycle Implementation.

The two outputs of the register file then feed into the data ALU inputs. The control units setup the ALU operation required to execute the instruction. Next, Load and Store instructions read or write to data memory. R-format instructions bypass data memory using a multiplexer. Last, R-format and Load instructions write back a new value into the register file.

PC-relative branch instructions use the adder and multiplexer shown above the data ALU in Figure 13.1 to compute the branch address. The multiplexer is required for conditional branch operations. After all outputs have stabilized, the next clock loads in the new value of the PC and the process repeats for the next instruction.

RISC instruction sets are easier to pipeline. With pipelining, the fetch, decode, execute, data memory, and register file write operations all work in parallel. In a single clock cycle, five different instructions are present in the pipeline. The basis for a pipelined hardware implementation of the MIPS is shown in Figure 13.2.

Additional complications arise because of data dependencies between instructions in the pipeline and branch operations. These problems can be resolved using hazard detection, data forwarding techniques, and branch flushing. With pipelining, most RISC instructions execute in one clock cycle. Branch instructions will still require flushing of the pipeline. Exercises that add pipelining to the processor core are included at the end of the chapter.

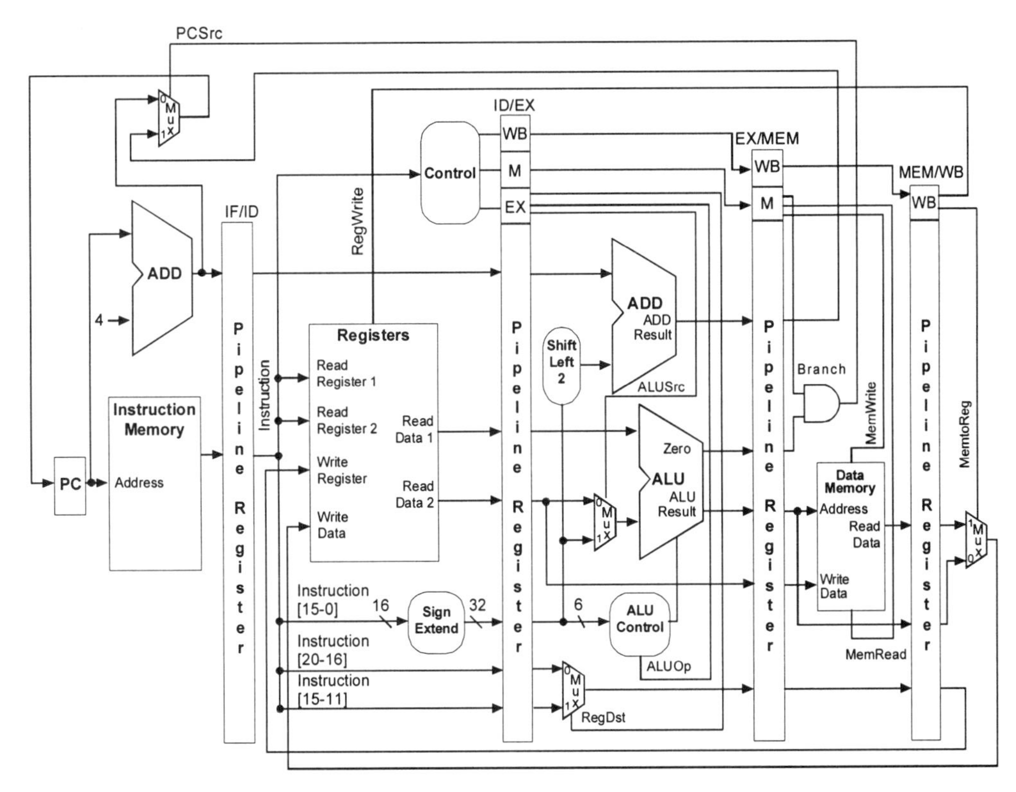

Figure 13.2 MIPS Pipelined Implementation.

13.2 Using VHDL to Synthesize the MIPS Processor Core

A VHDL-synthesis model of the MIPS single clock cycle model from Figure 13.1 will be developed in this section. This model can be used for simulation and implemented using the UP 1 board.

A full 32-bit version of the MIPS requires several minutes to synthesize on the PC platform and will not fit on the UP 1 board's FLEX 10K20 CPLD. A smaller version containing an 8-bit data path with full 32-bit instructions and control was developed to permit the design to synthesize and simulate in just a few minutes for laboratory exercises. This version has minimal VHDL source level modifications from the 32-bit version. Changes are limited to the bit vector array size declarations for variables and signals in the data path. If fast systems are available and long run times are not an issue, the width of the data path could be easily changed back to 32 bits. The smaller model requires machine language test programs that use 8-bit integer values. Since the hardware test programs needed are just a few instructions long and relatively simple, this does not impact the instructional value of the MIPS VHDL model.

A two-level hierarchy is used in the model. TOP_SPIM.VHD is the top-level of the hierarchy. It consists of a structural VHDL model that connects the five behavioral modules. The five behavioral modules are already setup so that they correspond to the different stages for the MIPS. This makes it much easier to modify when the model is pipelined in later laboratory exercises. For many synthesis tools, hierarchy is also required to synthesize large logic designs.

IFETCH.VHD is the VHDL submodule that contains instruction memory and the program counter. CONTROL.VHD contains the logic for the control unit. IDECODE.VHD contains the dual-ported register file. EXECUTE.VHD contains the data and branch address ALUs. DMEMORY.VHD contains the data memory.

13.3 The Top-Level Module

The TOP_SPIM.VHD file contains the top-level design file. TOP_SPIM is a VHDL structural model that connects the five component parts of the MIPS. This module could also be created using the schematic editor and connecting the symbols for each VHDL submodule. The inputs are the clock and reset signals. The values of major busses and important control signals are copied and output from the top level so that they are available for easy display in simulations.

```
                                    -- Top Level Structural Model, TOP_SPIM
LIBRARY IEEE;
USE IEEE.STD_LOGIC_1164.ALL;
USE IEEE.STD_LOGIC_ARITH.ALL;

ENTITY TOP_SPIM IS

   PORT( reset, clock                    : IN    STD_LOGIC;
      -- Output important signals to pins for easy display in Simulator
      PC, ALU_result_out, read_data_1_out,
      read_data_2_out, write_data_out    : OUT STD_LOGIC_VECTOR( 7 DOWNTO 0 );
      Instruction_out                    : OUT STD_LOGIC_VECTOR( 31 DOWNTO 0 );
      Branch_out, Zero_out, Memwrite_out,
      Regwrite_out                       : OUT STD_LOGIC );
END   TOP_SPIM;

ARCHITECTURE structure OF TOP_SPIM IS

   COMPONENT Ifetch
     PORT(  Instruction          : OUT   STD_LOGIC_VECTOR( 31 DOWNTO 0 );
            PC_plus_4_out        : OUT   STD_LOGIC_VECTOR( 7 DOWNTO 0 );
            Add_result           : IN    STD_LOGIC_VECTOR( 7 DOWNTO 0 );
            Branch               : IN    STD_LOGIC;
            Zero                 : IN    STD_LOGIC;
            PC_out               : OUT   STD_LOGIC_VECTOR( 7 DOWNTO 0 );
            clock,reset          : IN    STD_LOGIC );
```

```
END COMPONENT;

COMPONENT Idecode
   PORT(  read_data_1          : OUT  STD_LOGIC_VECTOR( 7 DOWNTO 0 );
          read_data_2          : OUT  STD_LOGIC_VECTOR( 7 DOWNTO 0 );
          Instruction          : IN   STD_LOGIC_VECTOR( 31 DOWNTO 0 );
          read_data            : IN   STD_LOGIC_VECTOR( 7 DOWNTO 0 );
          ALU_result           : IN   STD_LOGIC_VECTOR( 7 DOWNTO 0 );
          RegWrite, MemtoReg   : IN   STD_LOGIC;
          RegDst               : IN   STD_LOGIC;
          Sign_extend          : OUT  STD_LOGIC_VECTOR( 7 DOWNTO 0 );
          clock, reset         : IN   STD_LOGIC );
END COMPONENT;

COMPONENT control
   PORT(  Opcode               : IN   STD_LOGIC_VECTOR( 5 DOWNTO 0 );
          RegDst               : OUT  STD_LOGIC;
          ALUSrc               : OUT  STD_LOGIC;
          MemtoReg             : OUT  STD_LOGIC;
          RegWrite             : OUT  STD_LOGIC;
          MemRead              : OUT  STD_LOGIC;
          MemWrite             : OUT  STD_LOGIC;
          Branch               : OUT  STD_LOGIC;
          ALUop                : OUT  STD_LOGIC_VECTOR( 1 DOWNTO 0 );
          clock, reset         : IN   STD_LOGIC );
END COMPONENT;

COMPONENT  Execute
   PORT(  Read_data_1          : IN   STD_LOGIC_VECTOR( 7 DOWNTO 0 );
          Read_data_2          : IN   STD_LOGIC_VECTOR( 7 DOWNTO 0 );
          Sign_Extend          : IN   STD_LOGIC_VECTOR( 7 DOWNTO 0 );
          Function_opcode      : IN   STD_LOGIC_VECTOR( 5 DOWNTO 0 );
          ALUOp                : IN   STD_LOGIC_VECTOR( 1 DOWNTO 0 );
          ALUSrc               : IN   STD_LOGIC;
          Zero                 : OUT  STD_LOGIC;
          ALU_Result           : OUT  STD_LOGIC_VECTOR( 7 DOWNTO 0 );
          Add_Result           : OUT  STD_LOGIC_VECTOR( 7 DOWNTO 0 );
          PC_plus_4            : IN   STD_LOGIC_VECTOR( 7 DOWNTO 0 );
          clock, reset         : IN   STD_LOGIC );
END COMPONENT;


COMPONENT dmemory
   PORT(  read_data            : OUT  STD_LOGIC_VECTOR( 7 DOWNTO 0 );
          address              : IN   STD_LOGIC_VECTOR( 7 DOWNTO 0 );
          write_data           : IN   STD_LOGIC_VECTOR( 7 DOWNTO 0 );
          MemRead, Memwrite    : IN   STD_LOGIC;
          Clock,reset          : IN   STD_LOGIC );
END COMPONENT;

                          -- declare signals used to connect VHDL components
SIGNAL PC_plus_4               :  STD_LOGIC_VECTOR( 7 DOWNTO 0 );
```

```
    SIGNAL read_data_1          : STD_LOGIC_VECTOR( 7 DOWNTO 0 );
    SIGNAL read_data_2          : STD_LOGIC_VECTOR( 7 DOWNTO 0 );
    SIGNAL Sign_Extend          : STD_LOGIC_VECTOR( 7 DOWNTO 0 );
    SIGNAL Add_result           : STD_LOGIC_VECTOR( 7 DOWNTO 0 );
    SIGNAL ALU_result           : STD_LOGIC_VECTOR( 7 DOWNTO 0 );
    SIGNAL read_data            : STD_LOGIC_VECTOR( 7 DOWNTO 0 );
    SIGNAL ALUSrc               : STD_LOGIC;
    SIGNAL Branch               : STD_LOGIC;
    SIGNAL RegDst               : STD_LOGIC;
    SIGNAL Regwrite             : STD_LOGIC;
    SIGNAL Zero                 : STD_LOGIC;
    SIGNAL MemWrite             : STD_LOGIC;
    SIGNAL MemtoReg             : STD_LOGIC;
    SIGNAL MemRead              : STD_LOGIC;
    SIGNAL ALUop                : STD_LOGIC_VECTOR(  1 DOWNTO 0 );
    SIGNAL Instruction          : STD_LOGIC_VECTOR( 31 DOWNTO 0 );

BEGIN
                                -- copy important signals to output pins for easy
                                -- display in Simulator
  Instruction_out   <= Instruction;
  ALU_result_out    <= ALU_result;
  read_data_1_out   <= read_data_1;
  read_data_2_out   <= read_data_2;
  write_data_out    <= read_data WHEN MemtoReg = '1' ELSE ALU_result;
  Branch_out        <= Branch;
  Zero_out          <= Zero;
  RegWrite_out      <= RegWrite;
  MemWrite_out      <= MemWrite;
                                -- connect the 5 MIPS components
  IFE : Ifetch
    PORT MAP (  Instruction     => Instruction,
                PC_plus_4_out   => PC_plus_4,
                Add_result      => Add_result,
                Branch          => Branch,
                Zero            => Zero,
                PC_out          => PC,
                clock           => clock,
                reset           => reset );

  ID : Idecode
    PORT MAP (  read_data_1     => read_data_1,
                read_data_2     => read_data_2,
                Instruction     => Instruction,
                read_data       => read_data,
                ALU_result      => ALU_result,
                RegWrite        => RegWrite,
                MemtoReg        => MemtoReg,
                RegDst          => RegDst,
                Sign_extend     => Sign_extend,
                clock           => clock,
                reset           => reset );
```

```
CTL: control
  PORT MAP ( Opcode             => Instruction( 31 DOWNTO 26 ),
             RegDst             => RegDst,
             ALUSrc             => ALUSrc,
             MemtoReg           => MemtoReg,
             RegWrite           => RegWrite,
             MemRead            => MemRead,
             MemWrite           => MemWrite,
             Branch             => Branch,
             ALUop              => ALUop,
             clock              => clock,
             reset              => reset );

EXE: Execute
  PORT MAP ( Read_data_1        => read_data_1,
             Read_data_2        => read_data_2,
             Sign_extend        => Sign_extend,
             Function_opcode    => Instruction( 5 DOWNTO 0 ),
             ALUOp              => ALUop,
             ALUSrc             => ALUSrc,
             Zero               => Zero,
             ALU_Result         => ALU_Result,
             Add_Result         => Add_Result,
             PC_plus_4          => PC_plus_4,
             Clock              => clock,
             Reset              => reset );

MEM: dmemory
  PORT MAP ( read_data          => read_data,
             address            => ALU_Result,
             write_data         => read_data_2,
             MemRead            => MemRead,
             Memwrite           => MemWrite,
             clock              => clock,
             reset              => reset );
END structure;
```

13.4 The Control Unit

The control unit of the MIPS shown in Figure 13.3 examines the instruction opcode bits and generates eight control signals used by the other stages of the processor.

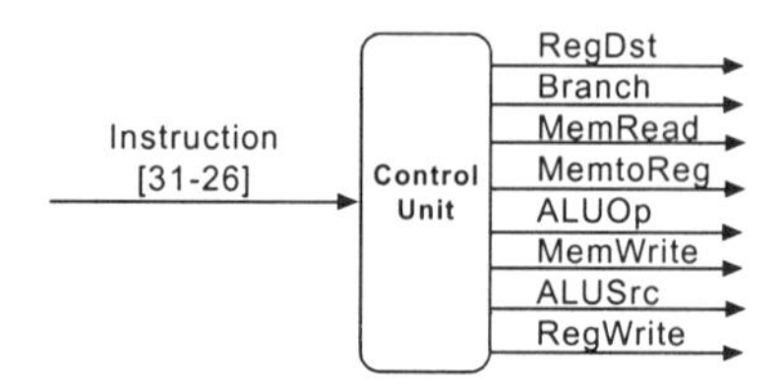

Figure 13.3 Block Diagram of MIPS Control Unit.

```
        -- control module (implements MIPS control unit)
LIBRARY IEEE;
USE IEEE.STD_LOGIC_1164.ALL;
USE IEEE.STD_LOGIC_ARITH.ALL;
USE IEEE.STD_LOGIC_SIGNED.ALL;
LIBRARY lpm;
USE lpm.lpm_components.ALL;

ENTITY control IS
  PORT(
    SIGNAL Opcode          : IN   STD_LOGIC_VECTOR( 5 DOWNTO 0 );
    SIGNAL RegDst          : OUT  STD_LOGIC;
    SIGNAL ALUSrc          : OUT  STD_LOGIC;
    SIGNAL MemtoReg        : OUT  STD_LOGIC;
    SIGNAL RegWrite        : OUT  STD_LOGIC;
    SIGNAL MemRead         : OUT  STD_LOGIC;
    SIGNAL MemWrite        : OUT  STD_LOGIC;
    SIGNAL Branch          : OUT  STD_LOGIC;
    SIGNAL ALUop           : OUT  STD_LOGIC_VECTOR( 1 DOWNTO 0 );
    SIGNAL clock, reset    : IN   STD_LOGIC );
END control;


ARCHITECTURE behavior OF control IS

COMPONENT LCELL
      PORT ( a_in    : IN STD_LOGIC;
             a_out   : OUT STD_LOGIC);
      END COMPONENT;

   SIGNAL  R_format, Lw, Sw, Beq     : STD_LOGIC;
   SIGNAL  Opcode_out                : STD_LOGIC_VECTOR( 5 DOWNTO 0 );

BEGIN
    Op_Buf0: LCELL  PORT MAP ( a_in  => Opcode( 0 ), a_out => Opcode_out( 0 ) );
    Op_Buf1: LCELL  PORT MAP ( a_in  => Opcode( 1 ), a_out => Opcode_out( 1 ) );
    Op_Buf2: LCELL  PORT MAP ( a_in  => Opcode( 2 ), a_out => Opcode_out( 2 ) );
    Op_Buf3: LCELL  PORT MAP ( a_in  => Opcode( 3 ), a_out => Opcode_out( 3 ) );
    Op_Buf4: LCELL  PORT MAP ( a_in  => Opcode( 4 ), a_out => Opcode_out( 4 ) );
    Op_Buf5: LCELL  PORT MAP ( a_in  => Opcode( 5 ), a_out => Opcode_out( 5 ) );
```

```
                    -- Code to generate control signals using opcode bits
R_format    <= '1' WHEN Opcode_out = "000000" ELSE '0';
Lw          <= '1' WHEN Opcode_out = "100011" ELSE '0';
Sw          <= '1' WHEN Opcode_out = "101011" ELSE '0';
Beq         <= '1' WHEN Opcode_out = "000100" ELSE '0';
RegDst      <= R_format;
ALUSrc      <= Lw OR Sw;
MemtoReg    <= Lw;
RegWrite    <= R_format OR Lw;
MemRead     <= Lw;
MemWrite    <= Sw;
Branch      <= Beq;
ALUOp( 1 )  <= R_format;
ALUOp( 0 )  <= Beq;

END behavior;
```

The LCELL function is needed to isolate logic blocks in the MIPS design. This enables MAX+PLUS to complete the logic minimization and synthesis step for the large and long logic paths present in the single cycle MIPS. The LCELL functions can be removed once the MIPS design is pipelined in exercises.

13.5 The Instruction Fetch Stage

The instruction fetch stage of the MIPS shown in Figure 13.4 contains the instruction memory, the program counter, and the hardware to increment the program counter to compute the next instruction address.

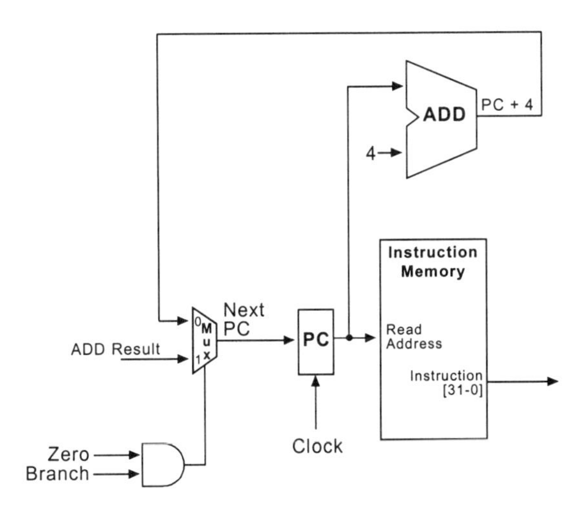

Figure 13.4 Block Diagram of MIPS Fetch Unit.

Instruction memory is implemented using the LPM_ROM function. 256 by 32 bits of instruction memory is available. This requires four of the FLEX chip's EAB memory blocks.

```
                                          -- Ifetch module (provides the PC and instruction
                                          --memory for the MIPS computer)
LIBRARY IEEE;
USE IEEE.STD_LOGIC_1164.ALL;
USE IEEE.STD_LOGIC_ARITH.ALL;
USE IEEE.STD_LOGIC_UNSIGNED.ALL;
LIBRARY lpm;
USE lpm.lpm_components.ALL;

ENTITY Ifetch IS
        PORT( SIGNAL Instruction        : OUT  STD_LOGIC_VECTOR( 31 DOWNTO 0 );
              SIGNAL PC_plus_4_out      : OUT  STD_LOGIC_VECTOR( 7 DOWNTO 0 );
              SIGNAL Add_result         : IN   STD_LOGIC_VECTOR( 7 DOWNTO 0 );
              SIGNAL Branch             : IN   STD_LOGIC;
              SIGNAL Zero               : IN   STD_LOGIC;
              SIGNAL PC_out             : OUT  STD_LOGIC_VECTOR( 7 DOWNTO 0 );
              SIGNAL clock, reset       : IN   STD_LOGIC);
END Ifetch;

ARCHITECTURE behavior OF Ifetch IS
        SIGNAL PC, PC_plus_4           : STD_LOGIC_VECTOR( 9 DOWNTO 0 );
        SIGNAL next_PC                 : STD_LOGIC_VECTOR( 7 DOWNTO 0 );
BEGIN
                                          --ROM for Instruction Memory
data_memory: lpm_rom

    GENERIC MAP (
        lpm_widthad          => 8,
        lpm_outdata          => "UNREGISTERED",
        lpm_address_control  => "UNREGISTERED",
                                  -- Reads in mif file for initial data memory values
        lpm_file             => "program.mif",
        lpm_width            => 32)
                                  -- Fetch next instruction from memory using PC
    PORT MAP (
        address         => PC( 9 DOWNTO 2 ),
        q               => Instruction );
                              -- copy output signals - allows read inside module
        PC_out          <= PC( 7 DOWNTO 0 );
        PC_plus_4_out <= PC_plus_4( 7 DOWNTO 0 );
                              -- Adder to increment PC by 4
        PC_plus_4( 9 DOWNTO 2 )  <= PC( 9 DOWNTO 2 ) + 1;
        PC_plus_4( 1 DOWNTO 0 )  <= "00";
                              -- Mux to select Branch Address or PC + 4
        next_PC  <= Add_result
                WHEN ( ( Branch = '1' ) AND ( Zero = '1' ) )
                ELSE  PC_plus_4( 9 DOWNTO 2 );
```

```
                                -- Store PC in register and load next PC on clock edge
    PROCESS
      BEGIN
        WAIT UNTIL ( clock'EVENT ) AND ( clock = '1' );
        IF reset = '1' THEN
              PC <= To_Stdlogicvector( B"0000000000" );
        ELSE
              PC( 9 DOWNTO 2 ) <= next_PC;
        END IF;
    END PROCESS;
END behavior;
```

The MIPS program is contained in instruction memory. Instruction memory is automatically initialized using the program.mif file shown in Figure 13.5. This initialization only occurs once during download and not at a reset. For different test programs, the appropriate machine code must be entered in this file in hex. Note that memory addresses displayed in the program.mif file are word addresses while addresses in registers such as the PC are byte addresses. The byte address is four times the word address since a word contains four bytes.

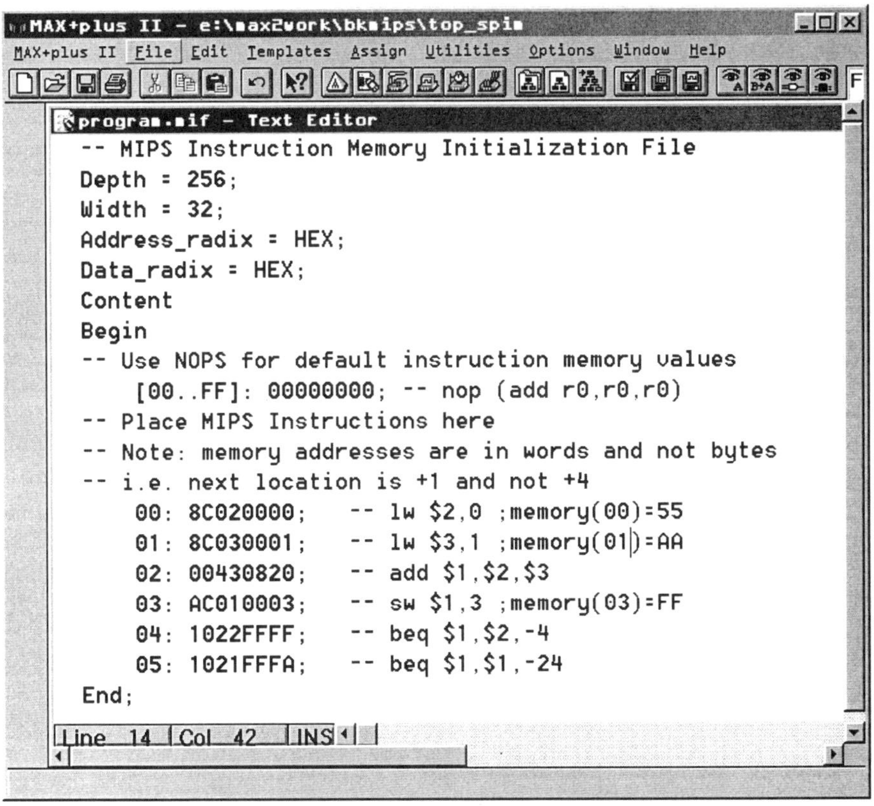

```
MAX+plus II - e:\max2work\bkmips\top_spim
MAX+plus II  File  Edit  Templates  Assign  Utilities  Options  Window  Help

program.mif - Text Editor
-- MIPS Instruction Memory Initialization File
Depth = 256;
Width = 32;
Address_radix = HEX;
Data_radix = HEX;
Content
Begin
-- Use NOPS for default instruction memory values
    [00..FF]: 00000000; -- nop (add r0,r0,r0)
-- Place MIPS Instructions here
-- Note: memory addresses are in words and not bytes
-- i.e. next location is +1 and not +4
    00: 8C020000;    -- lw $2,0 ;memory(00)=55
    01: 8C030001;    -- lw $3,1 ;memory(01)=AA
    02: 00430820;    -- add $1,$2,$3
    03: AC010003;    -- sw $1,3 ;memory(03)=FF
    04: 1022FFFF;    -- beq $1,$2,-4
    05: 1021FFFA;    -- beq $1,$1,-24
End;

Line 14  Col 42  INS
```

Figure 13.5 MIPS Program Memory Initialization File, program.mif.

13.6 The Decode Stage

The decode stage of the MIPS contains the register file as shown in Figure 13.6. The MIPS contains thirty-two 32-bit registers. To speed up synthesis and simulation, the register file has been reduced to eight 8-bit registers. R0 through R7 are implemented. The register file requires a major portion of the hardware required to implement the MIPS. Registers are initialized to the register number during a reset. This is done to enable the use of shorter test programs that do not have to load all of the registers. A VHDL FOR...LOOP structure is used to generate the initial register values at reset.

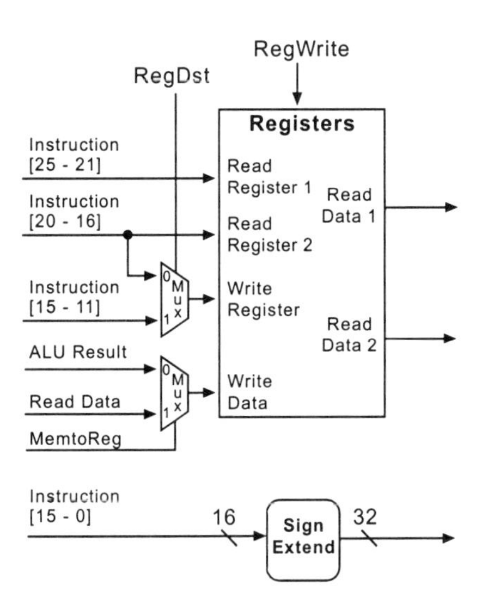

Figure 13.6 Block Diagram of MIPS Decode Unit.

```
                                -- Idecode module (implements the register file for
LIBRARY IEEE;                   -- the MIPS computer)
USE IEEE.STD_LOGIC_1164.ALL;
USE IEEE.STD_LOGIC_ARITH.ALL;
USE IEEE.STD_LOGIC_UNSIGNED.ALL;

ENTITY Idecode IS
    PORT(   read_data_1     : OUT   STD_LOGIC_VECTOR( 7 DOWNTO 0 );
            read_data_2     : OUT   STD_LOGIC_VECTOR( 7 DOWNTO 0 );
            Instruction     : IN    STD_LOGIC_VECTOR( 31 DOWNTO 0 );
            read_data       : IN    STD_LOGIC_VECTOR( 7 DOWNTO 0 );
            ALU_result      : IN    STD_LOGIC_VECTOR( 7 DOWNTO 0 );
            RegWrite        : IN    STD_LOGIC;
            MemtoReg        : IN    STD_LOGIC;
            RegDst          : IN    STD_LOGIC;
            Sign_extend     : OUT   STD_LOGIC_VECTOR( 7 DOWNTO 0 );
            clock,reset     : IN    STD_LOGIC );
END Idecode;
```

```
ARCHITECTURE behavior OF Idecode IS
TYPE register_file IS ARRAY ( 0 TO 7 ) OF STD_LOGIC_VECTOR( 7 DOWNTO 0 );

    SIGNAL register_array: register_file;
    SIGNAL write_register_address       : STD_LOGIC_VECTOR( 4 DOWNTO 0 );
    SIGNAL write_data                   : STD_LOGIC_VECTOR( 7 DOWNTO 0 );
    SIGNAL read_register_1_address      : STD_LOGIC_VECTOR( 4 DOWNTO 0 );
    SIGNAL read_register_2_address      : STD_LOGIC_VECTOR( 4 DOWNTO 0 );
    SIGNAL write_register_address_1     : STD_LOGIC_VECTOR( 4 DOWNTO 0 );
    SIGNAL write_register_address_0     : STD_LOGIC_VECTOR( 4 DOWNTO 0 );
    SIGNAL Instruction_immediate_value : STD_LOGIC_VECTOR( 15 DOWNTO 0 );

BEGIN
    read_register_1_address      <= Instruction( 25 DOWNTO 21 );
    read_register_2_address      <= Instruction( 20 DOWNTO 16 );
    write_register_address_1     <= Instruction( 15 DOWNTO 11 );
    write_register_address_0     <= Instruction( 20 DOWNTO 16 );
    Instruction_immediate_value  <= Instruction( 15 DOWNTO 0 );
                                  -- Read Register 1 Operation
    read_data_1 <= register_array(
                   CONV_INTEGER( read_register_1_address( 2 DOWNTO 0 ) ) );
                                  -- Read Register 2 Operation
    read_data_2 <= register_array(
                   CONV_INTEGER( read_register_2_address( 2 DOWNTO 0 ) ) );
                                  -- Mux for Register Write Address
    write_register_address <= write_register_address_1
              WHEN RegDst = '1'          ELSE write_register_address_0;
                                  -- Mux to bypass data memory for Rformat instructions
    write_data <= ALU_result( 7 DOWNTO 0 )
              WHEN ( MemtoReg = '0' )  ELSE read_data;

    Sign_extend <= Instruction_immediate_value( 7 DOWNTO 0 );

PROCESS
    BEGIN
        WAIT UNTIL clock'EVENT AND clock = '1';
        IF reset = '1' THEN
                                  -- Initial register values on reset are register = reg#
                                  -- use loop to automatically generate reset logic
                                  -- for all registers
            FOR i IN 0 TO 7 LOOP
                register_array(i) <= CONV_STD_LOGIC_VECTOR( i, 8 );
            END LOOP;
                                  -- Write back to register - don't write to register 0
        ELSIF RegWrite = '1' AND write_register_address /= 0 THEN
            register_array( CONV_INTEGER( write_register_address( 2 DOWNTO 0 ))) <=
                                                                    write_data;
        END IF;
    END PROCESS;
END behavior;
```

13.7 The Execute Stage

The execute stage of the MIPS shown in Figure 13.7 contains the data ALU and a branch address adder used for PC-relative branch instructions. Multiplexers that select different data for the ALU input are also in this stage. To speed synthesis and simulation, the width of the data ALU is limited to eight bits.

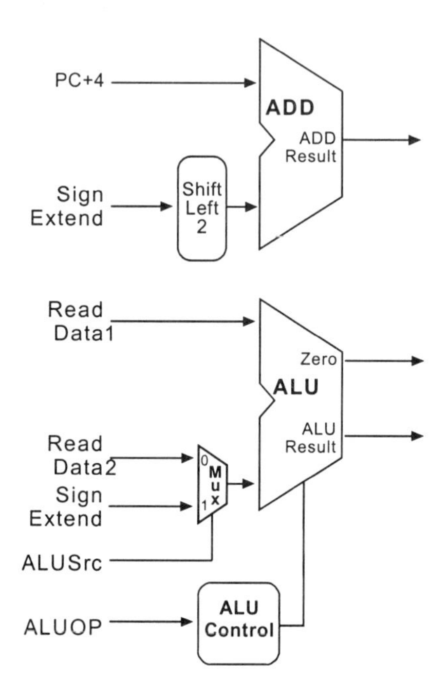

Figure 13.7 Block Diagram of MIPS Execute Unit.

```
-- Execute module (implements the data ALU and Branch Address Adder
-- for the MIPS computer)
LIBRARY IEEE;
USE IEEE.STD_LOGIC_1164.ALL;
USE IEEE.STD_LOGIC_ARITH.ALL;
USE IEEE.STD_LOGIC_SIGNED.ALL;

ENTITY  Execute IS
    PORT( Read_data_1     : IN    STD_LOGIC_VECTOR( 7 DOWNTO 0 );
          Read_data_2     : IN    STD_LOGIC_VECTOR( 7 DOWNTO 0 );
          Sign_extend     : IN    STD_LOGIC_VECTOR( 7 DOWNTO 0 );
          Function_opcode : IN    STD_LOGIC_VECTOR( 5 DOWNTO 0 );
          ALUOp           : IN    STD_LOGIC_VECTOR( 1 DOWNTO 0 );
          ALUSrc          : IN    STD_LOGIC;
          Zero            : OUT   STD_LOGIC;
          ALU_Result      : OUT   STD_LOGIC_VECTOR( 7 DOWNTO 0 );
          Add_Result      : OUT   STD_LOGIC_VECTOR( 7 DOWNTO 0 );
          PC_plus_4       : IN    STD_LOGIC_VECTOR( 7 DOWNTO 0 );
          clock, reset    : IN    STD_LOGIC );
END Execute;
```

```
ARCHITECTURE behavior OF Execute IS
SIGNAL Ainput, Binput           : STD_LOGIC_VECTOR( 7 DOWNTO 0 );
SIGNAL ALU_output_mux           : STD_LOGIC_VECTOR( 7 DOWNTO 0 );
SIGNAL Branch_Add               : STD_LOGIC_VECTOR( 8 DOWNTO 0 );
SIGNAL ALU_ctl                  : STD_LOGIC_VECTOR( 2 DOWNTO 0 );
BEGIN
   Ainput <= Read_data_1;
                                        -- ALU input mux
   Binput <= Read_data_2
      WHEN ( ALUSrc = '0' )
      ELSE  Sign_extend( 7 DOWNTO 0 );
                                        -- Generate ALU control bits
   ALU_ctl( 0 ) <= ( Function_opcode( 0 ) OR Function_opcode( 3 ) ) AND ALUOp(1 );
   ALU_ctl( 1 ) <= ( NOT Function_opcode( 2 ) ) OR (NOT ALUOp( 1 ) );
   ALU_ctl( 2 ) <= ( Function_opcode( 1 ) AND ALUOp( 1 )) OR ALUOp( 0 );
                                        -- Generate Zero Flag
   Zero <= '1'
      WHEN ( ALU_output_mux( 7 DOWNTO 0 ) = To_stdlogicvector( X"00" ) )
      ELSE '0';
                                        -- Select ALU output
   ALU_result <= To_stdlogicvector( B"0000000" ) & ALU_output_mux( 7 )
      WHEN  ALU_ctl = "111"
      ELSE  ALU_output_mux( 7 DOWNTO 0 );
                                        -- Adder to compute Branch Address
   Branch_Add <= PC_plus_4( 7 DOWNTO 2 ) +  Sign_extend( 7 DOWNTO 0 ) ;
   Add_result   <= Branch_Add( 7 DOWNTO 0 );

PROCESS ( ALU_ctl, Ainput, Binput )
   BEGIN
                              -- Select ALU operation
   CASE ALU_ctl IS
                                        -- ALU performs ALUresult = A_input AND B_input
      WHEN "000"     => ALU_output_mux   <= Ainput AND Binput;
                                        -- ALU performs ALUresult = A_input OR B_input
      WHEN "001"     => ALU_output_mux   <= Ainput OR Binput;
                                        -- ALU performs ALUresult = A_input + B_input
      WHEN "010"     => ALU_output_mux   <= Ainput + Binput;
                                        -- ALU performs SLT
      WHEN "011"     => ALU_output_mux   <= Ainput - Binput;
                                        -- ALU performs ?
      WHEN "100"     => ALU_output_mux   <= To_stdlogicvector( B"00000000" );
                                        -- ALU performs ?
      WHEN "101"     => ALU_output_mux   <= To_stdlogicvector( B"00000000" );
                                        -- ALU performs ALUresult = A_input -B_input
      WHEN "110"     => ALU_output_mux   <= Ainput - Binput;
                                        -- ALU performs SLT
      WHEN "111"     => ALU_output_mux   <= To_stdlogicvector( B"00000000" );
      WHEN OTHERS => ALU_output_mux   <= To_stdlogicvector( B"00000000" );
   END CASE;
 END PROCESS;
END behavior;
```

13.8 The Data Memory Stage

The data memory stage of the MIPS shown in Figure 13.8 contains the data memory. To speed synthesis and simulation, data memory is limited to 256 locations of 8-bit memory. Data memory is implemented using the LPM_RAM_DQ function. Memory write cycle timing is critical in any design. The LPM_RAM_DQ function requires that the memory address must be stable before write enable goes high. One FLEX EAB memory block is used for data memory. Four FLEX EABs are used for 32-bit instruction memory.

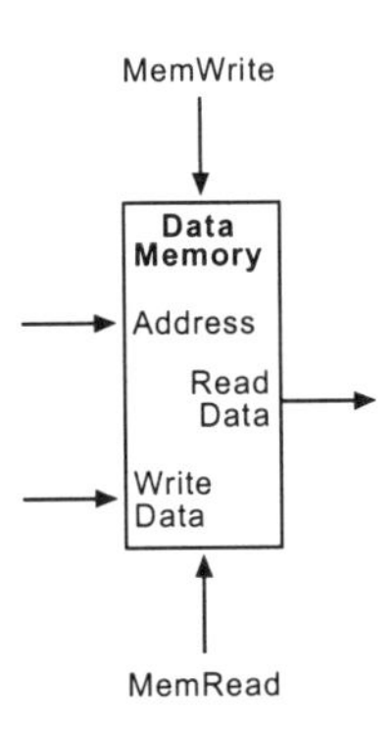

Figure 13.8 Block Diagram of MIPS Data Memory Unit.

```
-- Dmemory module (implements the data
-- memory for the MIPS computer)

LIBRARY IEEE;
USE IEEE.STD_LOGIC_1164.ALL;
USE IEEE.STD_LOGIC_ARITH.ALL;
USE IEEE.STD_LOGIC_SIGNED.ALL;
LIBRARY lpm;
USE lpm.lpm_components.ALL;

ENTITY dmemory IS
    PORT(   read_data           : OUT   STD_LOGIC_VECTOR( 7 DOWNTO 0 );
            address             : IN    STD_LOGIC_VECTOR( 7 DOWNTO 0 );
            write_data          : IN    STD_LOGIC_VECTOR( 7 DOWNTO 0 );
            MemRead, Memwrite   : IN    STD_LOGIC;
            clock,reset         : IN    STD_LOGIC );
END dmemory;

ARCHITECTURE behavior OF dmemory IS
SIGNAL lpm_write : STD_LOGIC;
BEGIN
```

```
data_memory: lpm_ram_dq
GENERIC MAP (
    lpm_widthad           => 8,
    lpm_outdata           => "UNREGISTERED",
    lpm_indata            => "REGISTERED",
    lpm_address_control   => "UNREGISTERED",
                          -- Reads in mif file for initial data memory values
    lpm_file              => "dmemory.mif",
    lpm_width             => 8 )

PORT MAP (
    data  => write_data, address => address,
    we    => lpm_write,  inclock  => clock,     q => read_data );
                          -- delay lpm write enable to ensure stable address and data
    lpm_write <= memwrite AND ( NOT clock );
END behavior;
```

MIPS data memory is initialized to the value specified in the file dmemory.mif shown in Figure 13.9. Note that the address displayed in the dmemory.mif file is a word address and not a byte address. Two values, 55 and AA, at byte address 0 and 4 are used for memory data in the short test program. The remaining locations are all initialized to zero.

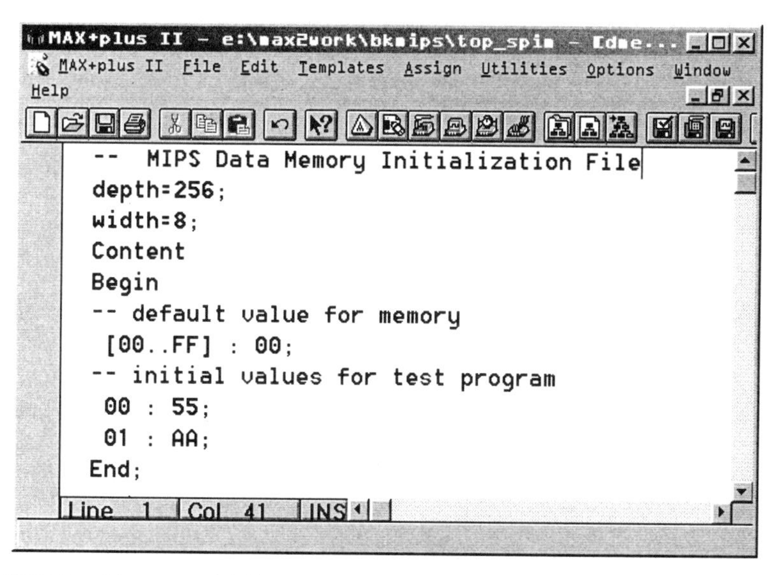

Figure 13.9 MIPS Data Memory Initialization File, dmemory.mif.

13.9 Simulation of the MIPS Design

The top-level file TOP_SPIM.VHD is compiled and used for simulation of the MIPS. It uses VHDL component instantiations to connect the five submodules. The values of major busses and important control signals are output at the top level for use in simulations. A reset is required to start the simulation with

PC = 0. A clock with a period of approximately 200ns is required for the simulation. Memory is initialized only at the start of the simulation. A reset does not re-initialize memory.

The execution of a short test program can be seen in the MIPS simulation output shown in Figure 13.10. The program loads two registers from memory with the LW instructions, adds the registers with an ADD, and stores the sum with SW. Next, the program does not take a BEQ conditional branch with a false branch condition. Last, the program loops back to the start of the program at PC = 0 with another BEQ conditional branch with a true branch condition.

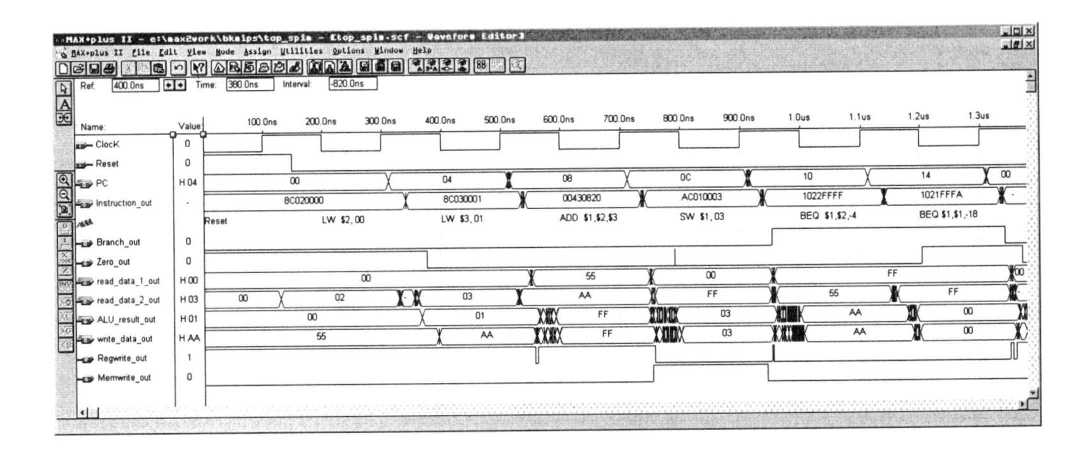

Figure 13.10 Simulation of MIPS test program.

13.10 MIPS Hardware Implementation on the UP 1 or UP 1X Board

A special version of the top level of the MIPS, TOP_FLEX.VHD, is identical to TOP_SPIM.VHD except that it also contains a VGA video output display driver. This driver displays the hexadecimal value of major busses in the MIPS processor. The video character generation technique used is discussed in Chapter 9. It also displays the PC on the seven-segment LED displays and uses the pushbuttons for clock and reset input. This top-level module should be used instead of TOP_SPIM.VHD after the design has been debugged in simulations. The final design with video output is then downloaded to the FLEX chip on the UP 1 or UP 1X board. The addition of the VGA video driver slows down the compile and simulation step, so it is faster to not add the video output while running initial simulations to debug a new design. The video driver uses three FLEX EAB memory blocks for character font data.

If the target device is a 10K20, additional EAB memory blocks are needed for video generation. An alternate implementation of MIPS instruction memory using a small array of constants is contained in the file, IFETCH2.VHD. This file uses LAB logic instead of EAB memory resources to implement a short instruction memory. This file must be used to replace IFETCH.VHD whenever the video driver is added to the MIPS design using a FLEX 10K20 CPLD.

If the larger FLEX 10K70 on the UP 2 or 1X board is used as the target device, this device contains the additional EAB memory needed without changes. A FLEX 10K70 or 10K50 can also be soldered onto the UP 1 board in place of the FLEX 10K20. Experience with surface mount soldering techniques is required. If surface mount soldering equipment is not available, it may be possible to find a local electronics assembly company that can replace the chip.

After simulation with TOP_SPIM.VHD, recompile using TOP_FLEX.VHD and download the design to the UP 1 board for hardware verification. Attach a VGA monitor to the UP 1's VGA connector. Any changes or additions made to signal names in TOP_SPIM.VHD and other modules will need to also be cut and pasted to TOP_FLEX.VHD.

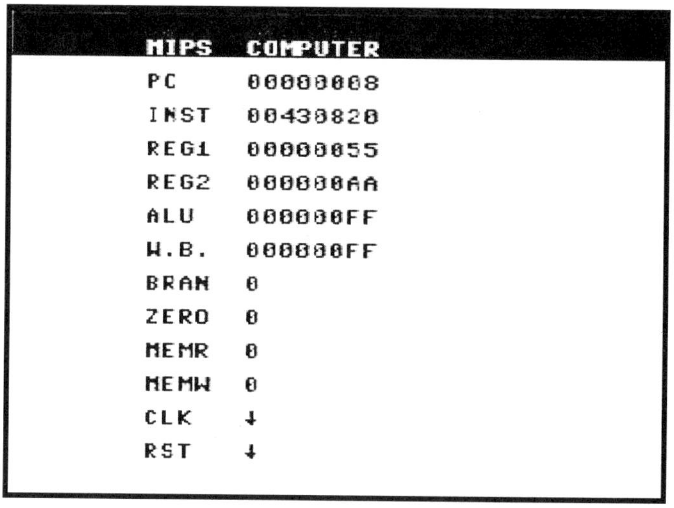

```
MIPS   COMPUTER
PC     00000008
INST   00430820
REG1   00000055
REG2   000000AA
ALU    000000FF
W.B.   000000FF
BRAN   0
ZERO   0
MEMR   0
MEMW   0
CLK    ↓
RST    ↓
```

Figure 13.11 MIPS with Video Output generated by UP 1 Board.

13.11 For Additional Information

The MIPS processor design and pipelining are described in the widely-used Patterson and Hennessy textbook, *Computer Organization and Design The Hardware/Software Interface*, Second Edition, Morgan Kaufman Publishers, 1998. The MIPS instructions are described in Chapter 3 and Appendix A of this text. The hardware design of the MIPS, used as the basis for this model, is described in Chapters 5 and 6 of the Patterson and Hennessy text.

SPIM, a free MIPS R2000 assembly language assembler and PC-based simulator developed by James Larus, is available free from http://www.cs.wisc.edu/~larus/spim.html . The reference manual for the SPIM simulator contains additional explanations of all of the MIPS instructions.

The MIPS instruction set and assembly language programming is also described in J. Waldron, *Introduction to RISC Assembly Language Programming*, Addison Wesley, 1999, and Kane and Heinrich, *MIPS RISC Architecture*, Prentice Hall, 1992.

13.12 Laboratory Exercises

1. Use VHDL to synthesize the MIPS single clock cycle design in the file TOP_SPIM.VHD. After synthesis and simulation perform the following steps:

   Display and print the timing diagram from the simulation. Verify that the information on the timing diagram shows that the hardware is functioning correctly. Examine the test program in IFETCH.VHD. Look at the program counter, the instruction bus, the register file and ALU outputs, and control signals on the timing diagram and carefully follow the execution of each instruction in the test program. Label the important values for each instruction on the timing diagram and attach a short write-up explaining in detail what the timing diagram shows relative to each instruction's execution and correct operation.

   Return to the simulator and run the simulation again. Examine the ALU output in the timing diagram window. Zoom in on the ALU output after execution of the fourth instruction and see what happens when it changes values. Explain exactly what is happening at this point. Hint: Real hardware has timing delays.

2. Recompile the MIPS model using the TOP_FLEX.VHD file, which generates video output. An alternate implementation of MIPS instruction memory using a small array of constants is contained in the file IFETCH2.VHD. This file uses LAB logic instead of EAB memory to implement a short instruction memory. This file must be used to replace IFETCH.VHD whenever the video driver is added to the MIPS design using a FLEX 10K20 CPLD. FLEX 10K50 or 10K70 devices have more EABs and do not require a different Ifetch module. Download the design to the UP 1 board. Attach a VGA monitor to the UP 1 board.

3. Write a MIPS test program for the AND, OR, and SUB instructions and run it on the VHDL MIPS simulation. These are all R-format instructions just like the ADD instruction. Modifications to the memory initialization files, program.mif and dmemory.mif, (i.e. only if you use data from memory in the test program) will be required. Registers have been preloaded with the register number to make it easy to run short test programs. Use registers $0..$7 for all test programs and recall that only the low eight bits are used.

4. Add and test the JMP instruction. The JMP or jump instruction is not PC-relative like the branch instructions. The J-format JMP instruction loads the PC with the low 26 bits of the instruction. Modifications to the existing VHDL MIPS model will be required. For a

suggested change, see the hardware modifications on page 306 of *Computer Organization and Design The Hardware/Software Interface.*

5. Add and test the BNE, branch if not equal, instruction. Modifications to the existing VHDL MIPS model will be required. Hint: Follow the implementation details of the existing BEQ, branch if equal, instruction and a change to add BNE should be obvious. Both BEQ and BNE must function correctly in a simulation. Be sure to test both the branch and no branch cases.

6. Add and test the I-format ADDIU, add immediate unsigned, instruction. Modifications to the existing VHDL MIPS model will be required.

7. Add and test the R-format SLT, set if less than, instruction. As an example SLT $1, $2, $3 performs the operation, If $2<$3 Then $1 = 1 Else $1 = 0. SLT is used before BEQ or BNE to implement the other branch conditions such as less than or greater.

8. Pipeline the MIPS VHDL simulation. Test your VHDL model by running a simulation of the example program shown in Figures 6.31-6.35 in *Computer Organization and Design The Hardware/Software Interface.* To minimize changes, pipeline registers must be placed in the VHDL module that generates the input to the pipeline. As an example, all of the pipeline registers that store control signals must be placed in the control module. Synthesize and check the control module first, since it is simple to see if it works correctly when you add the pipeline flip-flops. Use the following notation to add new latched signals, add a "D_" in front of the signal name to indicate it is the input to a D flip-flop used in a pipeline register. Signals that go through two D flip-flops would be "DD_" and three would be "DDD_". As an example, *instruction* would be the registered version of the signal, D_*instruction.*

   Add pipeline registers to the existing modules that generate the inputs to the pipeline registers shown in the text. This will prevent adding more modules and will not require extensive changes to the TOP_SPIM module. Add signal and process statements to model the pipeline modules – see the PC in the ifetch.vhd module for an example of how this can work. A few muxes may have to be moved to different modules.

   The control module should contain all of the control pipeline registers – 1, 2, or 3 stages of pipeline registers for control signals. Some control signals must be reset to zero, so use a D flip-flop with a synchronous reset for these pipeline registers. This generates a flip-flop with a Clear input that will be tied to Reset. Critical pipeline registers with control signals such as regwrite or memwrite should be cleared at reset so that the pipeline starts up correctly. The MIPS instruction ADD $0, $0, $0 is all zeros and does not modify any values in registers or memory. It is used to initialize the IF/ID pipeline at reset.

   Since only eight registers are implemented in the MIPS VHDL model, use the following modified test program:

```
Lw     $1,1($2)
Sub    $3,$4,$5
And    $2,$7,$6
Or     $4,$4,$5
Add    $5,$6,$7
```

Sections 6.2 and 6.3 of *Computer Organization and Design The Hardware/Software Interface* contain additional background information on pipelining.

9. Once the MIPS is pipelined as in problem 8, data hazards can occur between the five instructions present in the pipeline. As an example consider the following program:

```
Sub    $2,$1,$3
Add    $4,$2,$5
```

The subtract instruction stores a result in register 2 and the following add instruction uses register 2 as a source operand. The new value of register 2 is written into the register file by SUB $2,$1,$3 in the write-back stage after the old value of register 2 was read out by ADD $4,$2,$5 in the decode stage. This problem is fixed by adding two forwarding muxes to each ALU input in the execute stage. In addition to the existing values feeding in the two ALU inputs, the forwarding multiplexers can also select the last ALU result or the last value in the data memory stage. These muxes are controlled by comparing the rd, rt, and rs register address fields of instructions in the decode, execute, or data memory stages. Instruction rd fields will need to be added to the pipelines in the execute, data memory, and write-back stages for the forwarding compare operations. Since register 0 is always zero, do not forward register 0 values.

Add forwarding control to the pipelined model developed in problem 8. Test your VHDL model by running a simulation of the example program shown in Figures 6.41-42 of *Computer Organization and Design The Hardware/Software Interface* by Patterson and Hennessy.

Two forwarding multiplexers must also be added to the Idecode module so that a register file write and read to the same register work correctly in one clock cycle. If the register file write address equals one of the two read addresses, the register file write data value should be forwarded out the appropriate read data port instead of the normal register file data value.

Use the following test program that has been modified for eight registers:

```
Sub    $2,$1,$3
And    $4,$2,$5
Or     $3,$2,$6
Add    $5,$4,$2
Slt    $1,$6,$7
```

Sections 6.4 and 6.5 of *Computer Organization and Design The Hardware/Software Interface* contain additional background information on forwarding.

10. Add LW/SW forwarding to the pipelined model. This will allow an LW to be followed by an SW that uses the same register. One solution is suggested in *Computer Organization and Design The Hardware/Software Interface* by Patterson and Hennessy on page 488. Write a test program and verify correct operation in a simulation.

11. When a branch is taken, several of the instructions that follow a branch have already been loaded into the pipeline. A process called flushing is used to prevent the execution of these instructions. Several of the pipeline registers are cleared so that these instructions do not store any values to registers or memory or cause a forwarding operation. Add branch flushing to the pipelined MIPS VHDL model as shown in Figures 6.51 and 6.52 of the *Computer Organization and Design The Hardware/Software Interface* by Patterson and Hennessy. Note that two new forwarding multiplexers at the register file outputs (not shown in the Figure) are needed to eliminate the new Branch data hazards that appear when the branch comparator is moved to the decode stage. Use the following test program:

```
         Bne $0,$0,label1
         Add $1,$1,$2
         Add $2,$0,$3
label1:  Beq $1,$2,label2
         Add $1,$1,$1
         Add $2,$2,$2
         Nop
label2:  Beq $2,$1,label2
```

Section 6.6 of *Computer Organization and Design The Hardware/Software Interface* contains additional background information on branch hazards.

12. Use the timing analyzer to determine the maximum clock rate for the pipelined MIPS implementation, verify correct operation at this clock rate in a simulation, and compare the clock rate to the original non-pipelined MIPS implementation.

13. Redesign the pipelined MIPS VHDL model so that branch instructions have 1 delay slot (i.e. one instruction after the branch is executed even when the branch is taken). Rewrite the VHDL model of the MIPS and test the program from the previous problem assuming 1 delay slot. Move instructions around and add nops if needed.

14. Add the overflow exception hardware suggested at the end of Chapter 6 in Figures 6.55 and 6.56 of *Computer Organization and Design The Hardware/Software Interface* by Patterson and Hennessy. Add an overflow circuit that produces the exception with a test program containing an ADD instruction that overflows. Display the PC and the trap

address in your simulation. For test and simulation purposes make the exception address 6 instead of 40000040. Section 6.7 of *Computer Organization and Design The Hardware/Software Interface* contains additional background information on exceptions.

15. Add the required instructions to the model to run the MIPS bubble sort program from Chapter 3 of *Computer Organization and Design The Hardware/Software Interface.* After verifying correct operation with a simulation, download the design to the UP 1 board and trace execution of the program using the video output. Sort this four element array 4, 3, 5, 1.

16. Add programmed keyboard input and video output to the sort program from the previous problem using the keyboard, vga_sync, and char_rom UP1cores. Use a dedicated memory location to interface to I/O devices. Appendix A.36-38 of *Computer Organization and Design The Hardware/Software Interface* contains an explanation of MIPS memory-mapped terminal I/O.

17. Develop a VHDL synthesis model for another RISC processor's instruction set. Possible choices include the SUN SPARC, the DEC ALPHA, and the HP PARISC. Earlier hardware implementations of these processors designed before they became superscalar are more likely to fit on the FLEX 10K20 chip. If possible, use a UP 1X board equipped with a larger 10K70 device since it provides more hardware resources.

Appendix A: Generation of Pseudo Random Binary Sequences

In many applications, random sequences of binary numbers are needed. These applications include random number generation for games, automatic test pattern generation, data encryption and decryption, data compression, and data error coding. Since a properly operating digital circuit never produces a random result, these sequences are called pseudo random. A long pseudo-random binary sequence appears to be random at first glance.

Table A.1 shows how to make an "n" bit pseudo-random binary sequence generator. Here is how it works for n = 8. Build an 8-bit shift register that shifts left one bit per clock cycle. Find the entry in Table A.1 for n = 8. This entry shows that bits 8,6,5,4 should all XORed or XNORed together to generate the next shift input used as the low bit in the shift register. Recall that the order of XOR operations does not matter. Note that the low-bit number is "1" and not "0" in this table.

A state machine that is actually a non-binary counter is produced. The counter visits all 2^n-1 non-zero states once in a pseudo-random order and then repeats. Since the counter visits every state once, a uniform distribution of numbers from 1 to 2^n-1 is generated. In addition to a shift register, only a minimal number of XOR or XNOR gates are needed for the circuit. The circuit is easy to synthesize in a HDL such as VHDL since only a few lines are required to shift and XOR the appropriate bits. Note that the next value in the random sequence is actually 2x or 2x + 1 the previous value, x. For applications that may require a more truly random appearing sequence order, use a larger random sequence generator and select a disjoint subset of the bits and shuffle the output bits.

The initial value loaded in the counter is called the seed. The seed or the random number is never zero in this circuit. If a seed of zero is ever loaded in the shift register it will stay stuck at zero. If needed, the circuit can be modified so that it generates 2^n states. For the same initial seed value, the circuit will always generate the same sequence of numbers. In applications that wait for input, a random seed can be obtained by building a counter with a fast clock and saving the value of the counter for the seed when an input device such as a pushbutton is activated.

Additional information on pseudo-random binary sequence generators can be found in *HDL Chip Design* by D.J. Smith, Doone Publications, 1996, and Xilinx Application Note 52, 1996.

Table A.1 Primitive Polynomials Modulo 2.

n	XOR from bits	n	XOR from bits	n	XOR from bits	n	XOR from bits
3	3,2	45	45,44,42,41	87	87,74	129	129,124
4	4,3	46	46,45,26,25	88	88,87,17,16	130	130,127
5	5,3	47	47,42	89	89,51	131	131,130,84,83
6	6,5	48	48,47,21,20	90	90,89,72,71	132	132,103
7	7,6	49	49,40	91	91,90,8,7	133	133,132,82,81
8	8,6,5,4	50	50,49,24,23	92	92,91,80,79	134	134,77
9	9,5	51	51,50,36,35	93	93,91	135	135,124
10	10,7	52	52,49	94	94,73	136	136,135,11,10
11	11,9	53	53,52,38,37	95	95,84	137	137,116
12	12,6,4,1	54	54,53,18,17	96	96,94,49,47	138	138,137,131,130
13	13,4,3,1	55	55,31	97	97,91	139	139,136,134,131
14	14,5,3,1	56	56,55,35,34	98	98,87	140	140,111
15	15,14	57	57,50	99	99,97,54,52	141	141,140,110,109
16	16,15,13,4	58	58,39	100	100,63	142	142,121
17	17,14	59	59,58,38,37	101	101,100,95,94	143	143,142,123,122
18	18,11	60	60,59	102	102,101,36,35	144	144,143,75,74
19	19,6,2,1	61	61,60,46,45	103	103,94	145	145,93
20	20,17	62	62,61,6,5	104	104,103,94,93	146	146,145,87,86
21	21,19	63	63,62	105	105,89	147	147,146,110,109
22	22,21	64	64,63,61,60	106	106,91	148	148,121
23	23,18	65	65,47	107	107,105,44,42	149	149,148,40,39
24	24,23,22,17	66	66,65,57,56	108	108,77	150	150,97
25	25,22	67	67,66,58,57	109	109,108,103,102	151	151,148
26	26,6,2,1	68	68,59	110	110,109,98,97	152	152,151,87,86
27	27,5,2,1	69	69,67,42,40	111	111,101	153	153,152
28	28,25	70	70,69,55,54	112	112,110,69,67	154	154,152,27,25
29	29,27	71	71,65	113	113,104	155	155,154,124,123
30	30,6,4,1	72	72,66,25,19	114	114,113,33,32	156	156,155,41,40
31	31,28	73	73,48	115	115,114,101,100	157	157,156,131,130
32	32,22,2,1	74	74,73,59,58	116	116,115,46,45	158	158,157,132,131
33	33,20	75	75,74,65,64	117	117,115,99,97	159	159,128
34	34,27,2,1	76	76,75,41,40	118	118,85	160	160,159,142,141
35	35,33	77	77,76,47,46	119	119,111	161	161,143
36	36,25	78	78,77,59,58	120	120,113,9,2	162	162,161,75,74
37	37,5,4,3,2,1	79	79,70	121	121,103	163	163,162,104,103
38	38,6,5,1	80	80,79,43,42	122	122,121,63,62	164	164,163,151,150
39	39,35	81	81,77	123	123,121	165	165,164,135,134
40	40,38,21,19	82	82,79,47,44	124	124,87	166	166,165,128,127
41	41,38	83	83,82,38,37	125	125,124,18,17	167	167,161
42	42,41,20,19	84	84,71	126	126,125,90,89	168	168,166,153,151
43	43,42,38,37	85	85,84,58,57	127	127,126		
44	44,43,18,17	86	86,85,74,73	128	128,126,101,99		

Appendix B: MAX+PLUS II Design and Data File Extensions

Design Files

Altera Design File (*.adf)
EDIF Input File (*.edf)
Graphic Design File (*.gdf)
OrCAD Schematic File (*.sch)
State Machine File (*.smf)
Text Design File (*.tdf)
Verilog Design File (*.v)
VHDL Design File (*.vhd)
Waveform Design File (*.wdf)
Xilinx Netlist File (*.xnf)

Ancillary Data Files

Assignment and Configuration File (*.acf)
Assignment and Configuration Output (*.aco)
Command File (*.cmd)
EDIF Command File (*.edc)
Fit File (*.fit)
Intel Hexadecimal Format File (*.hex)
History File (*.hst)
Include File (*.inc)
JTAG chain file (*.jcf)
Library Mapping File (*.lmf)
Log File (*.log)
Memory Initialization File (*.mif)
Memory Initialization Output File (*.mio)
Message Text File (*.mtf)
Programmer Log File (*.plf)
Report File (*.rpt)
Simulator Channel File (*.scf)
Standard Delay Format (*.sdf)
Standard Delay Format Output File (*.sdo)
Symbol File (*.sym)
Table File (*.tbl)
Tabular Text File (*.ttf)
Text Design Export File (*.tdx)
Text Design Output File (*.tdo)
Timing Anlyzer Output File (*.tao)
Vector File (*.vec)
VHDL Memory Model Output File (*.vmo)

Non-Editable Ancillary File Types

Compiler Netlist File (*.cnf)
Hierarchy Interconnect File (*.hif)
JEDEC file (*.jed)
Node Database File (*.ndb)
Programmer Object File (*.pof)
Raw Binary File (*.rbf)
Serial Bitstream File (*.sbf)
Simulator Initialization File (*.sif)
Simulator Netlist File (*.snf)
SRAM Object File (*.sof)

Appendix C: UP 1 and UP 1X Pin Assignments

Table C.1 UP 1 and UP 1X board EPM7128S MAX CHIP I/O pin assignments.

Pin Name	Pin Type	Pin	Function of Pin
MSD_dp	OUTPUT PIN	68	Most Significant Digit of Seven-segment Display Decimal Point Segment (0 = LED ON, 1 = LED OFF)
MSD_g	OUTPUT PIN	67	MSD Display Segment G (0 = LED ON, 1 = LED OFF)
MSD_f	OUTPUT PIN	65	MSD Display Segment F (0 = LED ON, 1 = LED OFF)
MSD_e	OUTPUT PIN	64	MSD Display Segment E (0 = LED ON, 1 = LED OFF)
MSD_d	OUTPUT PIN	63	MSD Display Segment D (0 = LED ON, 1 = LED OFF)
MSD_c	OUTPUT PIN	61	MSD Display Segment C (0 = LED ON, 1 = LED OFF)
MSD_b	OUTPUT PIN	60	MSD Display Segment B (0 = LED ON, 1 = LED OFF)
MSD_a	OUTPUT PIN	58	MSD Display Segment A (0 = LED ON, 1 = LED OFF)
LSD_dp	OUTPUT PIN	79	Least Significant Digit of Seven-segment Display Decimal Point Segment (0 = LED ON, 1 = LED OFF)
LSD_g	OUTPUT PIN	77	LSD Display Segment G (0 = LED ON, 1 = LED OFF)
LSD_f	OUTPUT PIN	75	LSD Display Segment F (0 = LED ON, 1 = LED OFF)
LSD_e	OUTPUT PIN	76	LSD Display Segment E (0 = LED ON, 1 = LED OFF)
LSD_d	OUTPUT PIN	74	LSD Display Segment D (0 = LED ON, 1 = LED OFF)
LSD_c	OUTPUT PIN	73	LSD Display Segment C (0 = LED ON, 1 = LED OFF)
LSD_b	OUTPUT PIN	70	LSD Display Segment B (0 = LED ON, 1 = LED OFF)
LSD_a	OUTPUT PIN	69	LSD Display Segment A (0 = LED ON, 1 = LED OFF)
PB1	INPUT PIN	*	Pushbutton 1 (non–debounced, 0 = button depressed)
PB2	INPUT PIN	*	Pushbutton 2 (non–debounced, 0 = button depressed)
D1..D16 LEDs	OUTPUT PIN	*	16 Discrete LEDs - D1...D16 (0 = LED ON, 1 = LED OFF)
SW1 & SW2	INPUT PIN	*	MAX DIP Switch Inputs - SW*x*S1...SW*x*S8 (1 = Open, 0 = Closed)
Clock	INPUT PIN	83	25.175Mhz System Clock on low skew Global Clock Line
Prototyping Header Pins	INPUT, OUTPUT	1-84	Black Prototyping Headers next to MAX chip – Numbers are silk-screened on board. Pins 12, 33, 54, 75, and 83 are not available.
* Jumper wires from the switch or LED to the MAX prototyping headers are required to use these devices. Any available unused MAX header pin can be assigned to this device.			

Table C.2 UP 1 and UP 1X Board FLEX CHIP I/O pin assignments.

Pin Name	Pin Type	Pin	Function of Pin
MSD_dp	OUTPUT PIN	14	Most Significant Digit of Seven-segment Display - Decimal Point Segment (0 = LED ON, 1 = LED OFF)
MSD_g	OUTPUT PIN	13	MSD Display Segment G (0 = LED ON,1 = LED OFF)
MSD_f	OUTPUT PIN	12	MSD Display Segment F (0 = LED ON, 1 = LED OFF)
MSD_e	OUTPUT PIN	11	MSD Display Segment E (0 = LED ON, 1 = LED OFF)
MSD_d	OUTPUT PIN	9	MSD Display Segment D (0 = LED ON, 1 = LED OFF)
MSD_c	OUTPUT PIN	8	MSD Display Segment C (0 = LED ON, 1 = LED OFF)
MSD_b	OUTPUT PIN	7	MSD Display Segment B (0 = LED ON, 1 = LED OFF)
MSD_a	OUTPUT PIN	6	MSD Display Segment A (0 = LED ON, 1 = LED OFF)
LSD_dp	OUTPUT PIN	25	Least Significant Digit of Seven-segment Display - Decimal Point Segment (0 = LED ON, 1 = LED OFF)
LSD_g	OUTPUT PIN	24	LSD Display Segment G (0 = LED ON, 1 = LED OFF)
LSD_f	OUTPUT PIN	23	LSD Display Segment F (0 = LED ON, 1 = LED OFF)
LSD_e	OUTPUT PIN	21	LSD Display Segment E (0 = LED ON, 1 = LED OFF)
LSD_d	OUTPUT PIN	20	LSD Display Segment D (0 = LED ON, 1 = LED OFF)
LSD_c	OUTPUT PIN	19	LSD Display Segment C (0 = LED ON, 1 = LED OFF)
LSD_b	OUTPUT PIN	18	LSD Display Segment B (0 = LED ON, 1 = LED OFF)
LSD_a	OUTPUT PIN	17	LSD Display Segment A (0 = LED ON, 1 = LED OFF)
FLEX_switch_1	INPUT PIN	41	FLEX DIP Switch Input 1 (1 = Open, 0 = Closed)
FLEX_switch_2	INPUT PIN	40	FLEX DIP Switch Input 2 (1 = Open, 0 = Closed)
FLEX_switch_3	INPUT PIN	39	FLEX DIP Switch Input 3 (1 = Open, 0 = Closed)
FLEX_switch_4	INPUT PIN	38	FLEX DIP Switch Input 4 (1 = Open, 0 = Closed)
FLEX_switch_5	INPUT PIN	36	FLEX DIP Switch Input 5 (1 = Open, 0 = Closed)
FLEX_switch_6	INPUT PIN	35	FLEX DIP Switch Input 6 (1 = Open, 0 = Closed)
FLEX_switch_7	INPUT PIN	34	FLEX DIP Switch Input 7 (1 = Open, 0 = Closed)
FLEX_switch_8	INPUT PIN	33	FLEX DIP Switch Input 8 (1 = Open, 0 = Closed)
PB1	INPUT PIN	28	Pushbutton 1 (non–debounced, 0 = button depressed)
PB2	INPUT PIN	29	Pushbutton 2 (non–debounced, 0 = button depressed)
Horiz_Sync	OUTPUT PIN	240	VGA Video Signal - Horizontal Synchronization
Vert_Sync	OUTPUT PIN	239	VGA Video Signal - Vertical Synchronization
Blue	OUTPUT PIN	238	VGA Video Signal - Blue Video Data
Green	OUTPUT PIN	237	VGA Video Signal - Green Video Data
Red	OUTPUT PIN	236	VGA Video Signal - Red Video Data
PS2_CLK	BIDIRECTIONAL	30	Clock line for PS/2 Mouse and Keyboard
PS2_DATA	BIDIRECTIONAL	31	Data line for PS/2 Mouse and Keyboard
Clock	INPUT PIN	91	25.175 MHz System Clock on low skew Global Clock Line

Table C.3 FLEX Expansion B Header Pins.

Header Pin	FLEX Pin	Header Pin	FLEX Pin
1	Unregulated Power(7.2V)	2	GND
3	VCC (+5V)	4	GND
5	VCC (Right Servo VCC)	6	GND (Right Servo VCC)
7	No connection	8	DI1/99
9	DI2/92	10	DI3/210
11	DI4/212	12	DEV_CLR/209
13	DEV_OE/213	14	DEV_CLK2/211
15	109	16	110 (Right Servo Signal)
17	111	18	113
19	114	20	115
21	116	22	117
23	118	24	119
25	120	26	126
27	127	28	128
29	129	30	131
31	132	32	133
33	134	34	136
35	137	36	138
37	139	38	141
39	142	40	143
41	144	42	146
43	147	44	148
45	149	46	151
47	152	48	153(IR Sensor)
49	154 (IR Right LED)	50	156(Compass West)
51	157 (IR Left LED)	52	158(Compass East)
53	159	54	161(Compass South)
55	162 (Left Servo Signal)	56	163(Compass North)
57	VCC (Left Servo VCC)	58	GND (Left Servo GND)
59	VCC (+5V)	60	GND
Note: Pins in parenthesis are used for UP1-bot Interfacing			

Appendix D: The Wintim Meta Assembler

An assembler translates human readable symbolic assembly language programs into binary machine language that can then be loaded into the computers memory. A traditional assembler is developed for a particular machine whose register names, instruction mnemonics and machine instruction formats are completely defined. A meta assembler is a more flexible assembler that allows the user to define new instruction formats for a new machine. Once instruction formats and opcode mnemonics are defined by the user, the meta assembler then serves as an assembler. A meta assembler is useful for people that are designing a new computer, since they can use it to assemble programs for the new computer without writing a new assembler from scratch. WinTim is a meta assembler used to convert the symbolic strings of a source program to machine language code and to assign memory addresses to each machine instruction word and data storage location. There are two steps for meta assembly:

Definition or Meta Assembly Phase: The first step is to define all of the instruction formats and mnemonic names. A source file is written that contains all of the definitions. The definition file is processed by WinTIM and definition tables are produced for the assembly process. Only one definition file is needed assemble all of the assembly language programs for a given machine.

Assembly Phase: The second step is to assemble the assembly language program for the new instruction formats using the instruction definition tables produced in the definition phase. In this step, the meta assembler functions as a conventional assembler as it converts symbolic assembly language in a *.src file into binary machine language. A conventional style assembly listing is also produced along with an optional Altera format MIF file to initialize FPLD memory. Just as in a conventional assembler, there can be assembly errors.

An Example Using WinTIM for the simple computer

The first step is to develop the definition text file for the new computer design. Here is an example of the definition file for the simple computer design in chapter 8. An instruction for this computer has 16-bits words that consist of an 8-bit opcode and an 8-bit address field.

```
TITLE   ASSEMBLY LANGUAGE DEFINITION FILE FOR UP1 COMPUTER DESIGN
WORD    16
WIDTH   72
LINES   50
;*****************************************************************
; INSTRUCTION OPCODE LABELS - MUST BE 8-BITS, 2 Hex DIGITS
;*****************************************************************
LADD:   EQU     H#00
LSTORE: EQU     H#01
LLOAD:  EQU     H#02
LJUMP:  EQU     H#03
```

```
LJNEG: EQU      H#04
LSUB:  EQU      H#05
LXOR:  EQU      H#06
LOR:   EQU      H#07
LAND:  EQU      H#08
LJPOS: EQU      H#09
LZERO: EQU      H#0A
LADDI: EQU      H#0B
LSHL:  EQU      H#0C
LSHR:  EQU      H#0D
LIN:   EQU      H#0E
LOUT:  EQU      H#0F
LWAIT: EQU      H#10
;*****************************************************************
; DATA PSEUDO OPS
;*****************************************************************
;DB:         DEF       8VH#00      ;8-BIT DATA DIRECTIVE
DW:          DEF      16VH#0000   ;16-BIT DATA DIRECTIVE
;*****************************************************************
;ASSEMBLY LANGUAGE INSTRUCTIONS
;*****************************************************************
ADD:         DEF      LADD,8VH#00
STORE:       DEF      LSTORE,8VH#00
LOAD:        DEF      LLOAD,8VH#00
JUMP:        DEF      LJUMP,8VH#00
JNEG:        DEF      LJNEG,8VH#00
SUBT:        DEF      LSUB,8VH#00
XOR:         DEF      LXOR,8VH#00
OR:          DEF      LOR,8VH#00
AND:         DEF      LAND,8VH#00
JPOS:        DEF      LJPOS,8VH#00
ZERO:        DEF      LZERO,8VH#00
ADDI:        DEF      LADDI,8VH#00
SHL:         DEF      LSHL,H#0,4VH#0
SHR:         DEF      LSHR,H#0,4VH#0
IN:          DEF      LIN,8VH#00
OUT:         DEF      LOUT,8VH#00
WAIT:        DEF      LWAIT,8VH#00
END
```

All characters are normally uppercase. Examining the first lines of the definition file above, TITLE is a title string for each page of the listing file. WORD 16 defines the memory word size for the computer as 16-bits. WIDTH 72 sets the number of characters per line in the listing file to 72. LINES 50 sets the number of lines per page in the listing. All text after a ";" in a line is a comment.

The next few lines define the opcode values. LADD is a string that is defined to be equivalent (EQU) to 8-bits of zeros. Since the meta assembler must supply values to fill up fields containing a fixed number of bits, each bit value also has a bit length specified. H#00 means a two-digit hexadecimal value of all zeros and it would have a bit length of eight since each hex digit requires 4-bits. In addition to H#, the assembler supports B# for binary, D# for decimal, and Q# for octal bit values. The next few lines starting with L*xxx* define the remaining opcode values.

To declare and initialize words of memory for data storage the DW directive is created. DEF means define an instruction or in this case a word of data memory. The ":" is used to distinguish labels from directives like DEF and it is not part of the DW string. Labels typically start in column 1. In the DEF argument 16VH#0000, 16V instructs the assembler to place a 16-bit variable value in memory when the DW string is seen in the assembly source file. The value H#0000 specifies the 16-bit default value of zero. The default value is used if the argument for DW is not provided. Note that the proper number of bits should always be specified to avoid bit length errors during assembly

For each instruction, the mnemonic name and format must be defined using the DEF directive. The line ADD: DEF LADD, 8VH#00, instructs the assembler to emit a machine language instruction whenever the ADD string is seen in the source file. The 16-bit machine code has the high 8-bits set to the value of LADD, the add opcode (i.e. H#00) and the low 8-bits are set to the argument of the ADD instruction. The remaining instructions are now defined using additional DEF commands. Since this computer only has a single instruction format, the only difference in these lines is the instruction mnemonic name and the opcode value.

This definition file is then read by the meta assembler and is used to setup tables for the assembly process. It is possible to have syntax errors in the definition file. Any syntax errors must be corrected before the assembly step.

Now the assembly process can begin. Here is an assembly language program source file, *.src, for the computer. This short program is intended only to demonstrate assembler features and it does not compute anything useful.

```
TITLE  EXAMPLE UP1 COMPUTER ASSEMBLY LANGUAGE TEST PROGRAM
LIST F,W
LINES 50
;**********************************
; MACROS
;**********************************
ECHO:         MACRO   PORT
              IN      PORT
              OUT     PORT
              ENDM
;**********************************
; CONSTANTS
;**********************************
CON1:         EQU     2
DISPLAY:      EQU    H#00
SWITCH:       EQU    H#01
;**********************************
; PROGRAM AREA
;**********************************
ORG H#00
START:        LOAD LABEL1%:
```

```
                ADDI 1%:
                SHL  1
                SHR  CON1%:
                AND  H#0F
                OR   H#80
                SUBT LABEL2%:
                JPOS ENDP%:
                XOR  LABEL3%:
                ADD  (TABLE1 + 3)%:
                JNEG ENDP%:
                IN   SWITCH
                OUT  DISPLAY
; MACRO TEST
                ECHO H#10
                WAIT B#11000011
ENDP:           STORE LABEL1%:
LOOP:           JUMP LOOP%:
                JUMP START<%:
                JUMP $%:
;********************************
; DATA FOR TEST PROGRAM
;********************************
        ORG H#80
LABEL1:         DW H#0ACE
LABEL2:         DW H#0000
LABEL3:         DW H#FFFF  ;UNSIGNED LARGEST NUMBER
LABEL4:         DW H#7FFF  ;TWO'S COMPLEMENT LARGEST NUMBER
TABLE1:         DW H#0000
                DW H#0011
                DW H#0022
                DW H#0033
                DW H#0044
                DW H#0055
                DW H#0066
                DW H#0077
                DW H#0088
END
```

The first three lines setup the listing file titles and options. The next section defines a macro. Macros are like a text editor's substitute string command or C's #define feature. The macro example will be examined later. The next section sets up strings that can be used in place of constants in the assembly language program. The command ORG H#00 sets the origin to zero. This tells the assembler to start placing instructions at memory location zero.

The next several lines contain instructions. The line START: LOAD LABEL1%:, sets up a label, START, that has the value of address of the LOAD instruction. Labels are strings used to mark locations in an assembly language program. Not every instruction needs a label. If you need to branch to an instruction or refer to a data value, a label is typically used. Using the actual address value is a bad practice since any modifications to the program could easily change all of the addresses. The string LOAD is recognized as an instruction from the definition file, so the assembly process emits a LOAD machine instruction with the proper values. LABEL1 is a label that specifies the load address for the instruction. LABEL1 is defined at the start of the next section of code. The"%:" after label instructs the assembler to right justify ("%") and truncate (":") the extra bits of the label value so that it will fit into the 8-bit field in the instruction.

A few lines down, note that simple expressions involving labels such as (TABLE1 + 3)%: are also allowed. The line JUMP START<%: generates a PC relative address value. "<" is a special character to indicate PC relative. This means that the address stored in the instruction has the address of the instruction subtracted from the label value. Many computers have PC relative branch instructions since they save address bits. "$" is a special symbol in many assemblers that means the current address.

The final section sets up the variables for the program. An ORG statement is used to keep the data area away from the instruction area. Labels are used to identify variables and the DW directive is used to reserve a word in memory and define an initial value for memory. END is commonly used in assemblers to indicate the end of the source file.

The Macro ECHO is defined in the macro section at the beginning of the program. Macros are used like a substitute string command in a text editor. Macros must be defined before they are called. The macro is expanded into the text string defined between the MACRO and ENDM directive. Don't forget ENDM or the entire program will become a macro definition. In the case of this macro, whenever the string ECHO *number* is encountered in the source file it will be replaced with IN *number* and OUT *number. Number* is an argument to the macro. No arguments or multiple arguments separated by commas are also supported. All macros are expanded at the start of the assembly process prior to any other operation. Examine the listing file to see the Macro expansion.

The definition files and the assembly language source files can be created using any text editor. A sample assembly listing file is shown below.

```
Addr          Line EXAMPLE UP1 COMPUTER ASSEMBLY LANGUAGE TEST PROGRAM
                 1 TITLE EXAMPLE UP1 COMPUTER ASSEMBLY LANGUAGE TEST PROGRAM
                 2 LIST F,W
                 3 LINES 50
                 4 ;*********************************
                 5 ; MACROS
                 6 ;*********************************
                 7 ECHO:    MACRO    PORT
                 8          IN       PORT
                 9          OUT      PORT
                10          ENDM
                11 ;*********************************
                12 ; CONSTANTS
                13 ;*********************************
                14 CON1:        EQU   2
                15 DISPLAY:     EQU    H#00
                16 SWITCH:      EQU    H#01
                17 ;*********************************
```

```
                18 ; PROGRAM AREA
                19 ;********************************
00000           20             ORG H#00
00000 0280      21 START:      LOAD LABEL1%:
00001 0B01      22             ADDI 1%:
00002 0C01      23             SHL  1%:
00003 0D02      24             SHR  CON1%:
00004 080F      25             AND  H#0F
00005 0780      26             OR   H#80
00006 0581      27             SUBT LABEL2%:
00007 0910      28             JPOS ENDP%:
00008 0682      29             XOR  LABEL3%:
00009 0087      30             ADD  (TABLE1 + 3)%:
0000A 0410      31             JNEG ENDP%:
0000B 0E01      32             IN   SWITCH
0000C 0F00      33             OUT  DISPLAY
                34 ; MACRO TEST
                35             ECHO H#10
0000D 0E10      35 +         IN      H#10
0000E 0F10      35 +         OUT     H#10
                35 +         ENDM
0000F 10C3      36             WAIT B#11000011
00010 0180      37 ENDP:       STORE LABEL1%:
00011 0311      38 LOOP:       JUMP LOOP%:
00012 0300      39             JUMP START%:
00013 0313      40             JUMP $%:
                41 ;********************************
                42 ; DATA FOR TEST PROGRAM
                43 ;********************************
00080           44             ORG H#80
00080 0ACE      45 LABEL1:     DW H#0ACE
00081 0000      46 LABEL2:     DW H#0000
00082 FFFF      47 LABEL3:     DW H#FFFF     ;UNSIGNED LARGEST NUMBER
00083 7FFF      48 LABEL4:     DW H#7FFF     ;TWO'S COMPLEMENT LARGEST NUMBER
00084 0000      49 TABLE1:     DW H#0000
00085 0011      50             DW H#0011
00086 0022      51             DW H#0022
00087 0033      52             DW H#0033
00088 0044      53             DW H#0044
00089 0055      54             DW H#0055
0008A 0066      55             DW H#0066
0008B 0077      56             DW H#0077
0008C 0088      57             DW H#0088
                58 END
```

When developing a new definition file, the machine codes in the listing should always be verified. On lines that emit memory data, the first column of hex numbers in the listing file is the memory address and the second column is the machine instruction or memory data. Any errors in machine instructions will cause serious time-consuming problems later on when testing the computer.

WinTim Assembly Example

Here is a short example using WinTim to assemble the program for the simple computer design. Up1def.src is the definition file and Up1asm.src is the assembly language source file. First, install and start WinTIM. Open these two source files with **File⇨Open**. First select the **Up1def.src window** then **Assemble⇨Meta-assemble** to process the definition file. Next select the **Up1asm.src window** then **Assemble⇨Up1asm.src** to assemble the program. Then **Output⇨** both the **Listing** and **MIF file format** to see the assembled machine code in a listing. To use the machine code in the Altera tools in a MIF file, select the **MIF window** and then use **File ⇨Save**. Recall that MIF files can be automatically loaded into the FLEX chip's RAM.

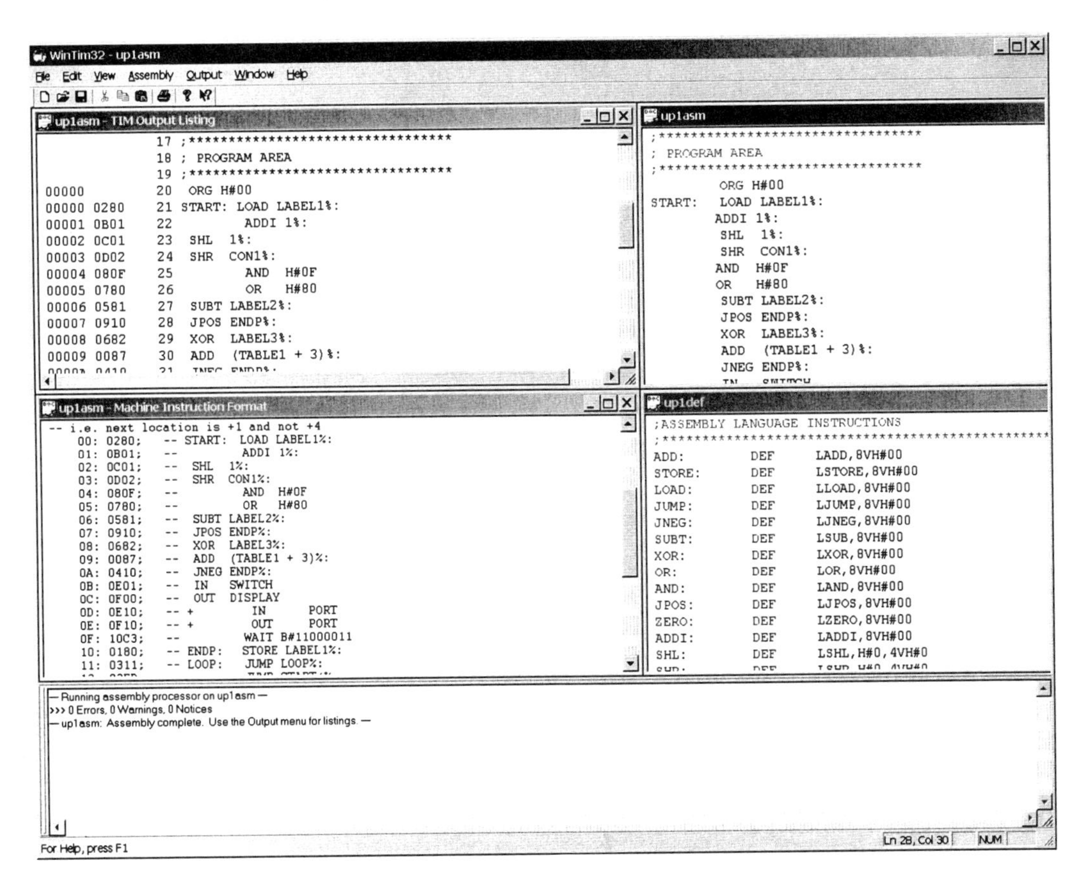

A more complex example using the MIPS computer

For a more complex instruction set consider the MIPS RISC processor. The MIPS has thirty-two registers and 32-bit instructions. In the definition file, it will be necessary to use equates to set up all of the register names and binary values. As an example for R4, the line R4: EQU B#00100 equates the string for register 4 to the five-bit binary value for four. Similar EQU lines will be needed for all of the remaining registers.

Next in the definition file, each of the instructions would be defined using a DEF command. The MIPS has only three instruction formats and all instructions are 32-bits. Only I-format LOAD and STORE instructions reference memory operands. R-format instructions such as ADD, AND, and OR perform operations only on data in the registers. They require two register operands, Rs and Rt. The result of the operation is stored in a third register, Rd. R-format shift and function fields are used as an extended opcode field. J-format instructions include the jump instructions.

Table E.1 MIPS 32-bit Instruction Formats.

Field Size	6-bits	5-bits	5-bits	5-bits	5-bits	6-bits
R- Format	Opcode	Rs	Rt	Rd	Shift	Function
I - Format	Opcode	Rs	Rt	Address/immediate value		
J - Format	Opcode	Branch target address				

R-Format instructions such as ADD could be defined as follows:

```
ADD: DEF  Q#00,5VB#00000,5VB#00000,5VB#00000,B#00000,B#100000
```

The Add instruction has an opcode of 0, three register arguments that default to R0, a shift field of 0, and a function code of 32. In assembly language, an example add instruction would appear as

```
ADD R1, R2, R3
```

The order of the three register arguments normally used in the MIPS assembly language is Rd, Rs, and Rt. This is not the same as the Rs, Rt, Rd bit order they are used in machine language. If desired, a macro can be used to swap the order of the arguments. Macros must have different names than instructions.

I-format Instructions such as LW could be defined as follows:

```
LW: DEF Q#43, 5VB#00000, 5VB#00000,16VH#0000
```

In the native MIPS assembler a typical LW example instruction would be represented as LW R1,Data_Label(R2). The optional parenthesis is used to specify the index register. In a meta assembler, it is often hard to create these special character meanings. A third argument is probably the best solution. In the meta assembler's assembly language, an example LW would be

```
LW R1,R2,Data_Label%:
```

In the native MIPS assembler, the instruction, LW R1,Label, specifies R0, which always is the value 0, as the index register. Note that if the argument is not specified in the meta assembler the default value of R0 will be used as in

```
LW R1,,Label%:
```

Once again the argument order could be swapped using a macro. Here is an example:

```
LW: MACRO ARG1, ARG2, ARG3
       ILW ARG1, ARG3, ARG2
    ENDM
```

Since the macro is now named LW, the instruction definition would need to be renamed as follows:

```
ILW: DEF Q#43, 5VB#00000, 5VB#00000,16VH#0000
```

The previous LW instruction example would now become

```
LW R1,Label%:
```

Conditional Assembly Directives

Experienced assembly language programmers often use conditional assembly directives. Conditional assembly directives can be used to automatically generate different versions of a program. Complex macros can use conditional assembly directives to generate different instruction sequences based on their arguments.

In WinTIM, the conditional assembly directive, IF *expression* ENDIF, is supported. If the *expression* is true, the following source lines until ENDIF are assembled, otherwise they are skipped. Note that *expression* must be evaluated at assembly time and not at run time. This means that *expression* cannot be a function of registers or other program variables that are defined only when the program runs on the computer. Here is an example using both macros and conditional assembly:

```
FIB:    MACRO   N
IF      N _NE_  0
ANS:    SET     ANS + N
        FIB     N-1
ENDIF
ENDM
```

The macro calls itself recursively with an argument of N-1 until N reaches zero. The conditional assembly directive skips the recursive call when N=0. SET is like EQU, but a symbol can be SET to different values during assembly. The next two lines initialize the ANS value and call the macro.

```
ANS:    SET     0
        FIB     5
```

After 4 recursive macro calls, ANS is 0+5+4+3+2+1=15 and the macro exits. Complex macros frequently examine the value of a constant argument to select different sequences of instructions.

This appendix has provided a brief overview of the features of Wintim. For further information or a more detailed explanation of any directive, refer to the extensive on-line help files provided with Wintim.

Appendix E: An Introduction to Verilog for VHDL users

VHDL and Verilog are the two most widely used hardware description languages. This appendix will present some simple Verilog synthesis examples so that users familiar with VHDL can understand the similar capabilities of Verilog. Table E.1 compares the operators used in VHDL and Verilog synthesis tools.

Table E.1 Comparison of VHDL and Verilog Operators.

<table>
<tr><th>Operator</th><th>VHDL</th><th>Verilog</th></tr>
<tr><td>Addition</td><td>+</td><td>+</td></tr>
<tr><td>Subtraction</td><td>-</td><td>-</td></tr>
<tr><td>Multiplication</td><td>*</td><td>*</td></tr>
<tr><td>Division</td><td>/</td><td>/</td></tr>
<tr><td>Modulus</td><td>MOD</td><td>%</td></tr>
<tr><td>Remainder</td><td>REM</td><td></td></tr>
<tr><td>Concatenation – used to combine bits</td><td>&</td><td>{ }</td></tr>
<tr><td>logical shift left</td><td>SLL**</td><td><<</td></tr>
<tr><td>logical shift right</td><td>SRL**</td><td>>></td></tr>
<tr><td>arithmetic shift left</td><td>SLA**</td><td></td></tr>
<tr><td>arithmetic shift right</td><td>SRA**</td><td></td></tr>
<tr><td>rotate left</td><td>ROL**</td><td></td></tr>
<tr><td>rotate right</td><td>ROR**</td><td></td></tr>
<tr><td>Equality</td><td>=</td><td>==</td></tr>
<tr><td>Inequality</td><td>/=</td><td>!=</td></tr>
<tr><td>less than</td><td><</td><td><</td></tr>
<tr><td>less than or equal</td><td><=</td><td><=</td></tr>
<tr><td>greater than</td><td>></td><td>></td></tr>
<tr><td>greater than or equal</td><td>>=</td><td>>=</td></tr>
<tr><td>logical NOT</td><td>NOT</td><td>!</td></tr>
<tr><td>logical AND</td><td>AND</td><td>&&</td></tr>
<tr><td>logical OR</td><td>OR</td><td>||</td></tr>
<tr><td>bitwise NOT</td><td>NOT</td><td>~</td></tr>
<tr><td>bitwise AND</td><td>AND</td><td>&</td></tr>
<tr><td>bitwise OR</td><td>OR</td><td>|</td></tr>
<tr><td>bitwise XOR</td><td>XOR</td><td>^</td></tr>
</table>

An example of a state machine design in both VHDL and Verilog can be seen in Figure E.1. A VHDL entity is equivalent to a Verilog module. In the first few lines of each model, input and output signals are declared.

VHDL State Machine Model

```
entity state_mach is
  port(clk, reset      : in  std_logic;
      input1, input2  : in  std_logic;
      Output1         : out std_logic);
end state_mach;
architecture A of state_mach is
  type STATE_TYPE is (state_A, state_B, state_C);
  signal state: STATE_TYPE;
begin
  process (reset, clk)
    begin
      if reset = '1' then
        state <= state_A;
      elsif clk'EVENT and clk = '1' then
        case state is
          when state_A =>
              if input1 = '0' then
                state <= state_B;
              else
                state <= state_C;
              end if;
            when state_B =>
              state <= state_C;
            when state_C =>
              if input2 = '1' then
                state <= state_A;
              end if;
        end case;
      end if;
  end process;
  with state select
    output1 <= '0' when state_A,
               '0' when state_B,
               '1' when state_C,
               '0' when others;
end a;
```

Verilog State Machine Model

```
module state_mach (clk, reset, input1,
             input2 ,output1);
  input clk, reset, input1, input2;
  output output1;
  reg output1;
  reg [1:0] state;
  parameter [1:0] state_A = 0, state_B = 1,
             state_C = 2;
always@(posedge clk or posedge reset)
  begin
    if (reset)
      state = state_A;
    else
      case (state)
        state_A:
          if (input1==0)
            state = state_B;
          else
            state = state_C;
        state_B:
          state = state_C;
        state_C:
          if (input2)
            state = state_A;
      endcase
  end
always @(state)
  begin
    case (state)
      state_A: output1 = 0;
      state_B: output1 = 0;
      state_C: output1 = 1;
      default:  output1 = 0;
    endcase
  end
endmodule
```

State Diagram

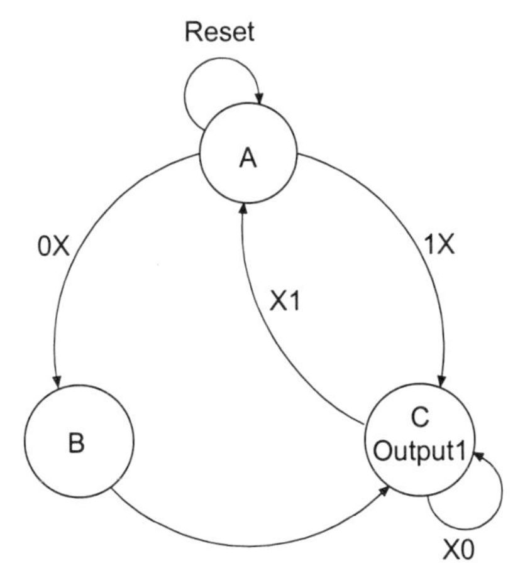

Figure E.1 Comparison of a state machine model in VHDL and Verilog

In Verilog, the reg statement is used to declare internal signal names. Declaring a signal with a reg statement does not necessarily mean that the signal will be held in a register after synthesis. A reg statement is used to declare internal signals that will be assigned in procedural assignment statements. Enumerated data types are not supported in Verilog, so it is necessary to assign a number to each state name with the parameter statement.

VHDL's processes are similar to Verilog's always blocks. Inside a Verilog always block, statements execute in sequential order just like a VHDL process. The statement always@(posedge clk) is similar to VHDL's clk'EVENT and clk='1' or rising_edge(clk). Sequential assignment statements inside an always@(posedge clk) block will be held in registers built using D flip-flops.

VHDL and Verilog both have concurrent statements and sequential statements. Inside an always block, sequential case and if statements can be used. Note that the control structure inside the first VHDL process and Verilog always block is very similar. The present value of state is examined in a case statement along with input signals to determine the next value of state. The second Verilog always block generates combinational logic to produce the state machine's output signal based on the current value of state. Note that the second always block specifies a sensitivity to the value of state with always@(state). Since there is no sensitivity to a clock signal no flip-flops are used for synthesis in this block.

The second example is a 16-bit ALU. Both VHDL and Verilog synthesis models are shown in Figure E.2. The output of the ALU feeds into a shift register and is then held in a register. The high two-bits of ALU_control select one of four ALU operations and the low-bit controls the shift operation. Once again the VHDL entity is a Verilog module and the two VHDL processes become two always blocks in Verilog. Different operators are used for bitwise AND, bitwise OR and shift in Verilog. Consult Table E.1 for the meaning of the new Verilog operators appearing in the case statement. The first always block generates the combinational logic for the ALU operations and the second synthesizes the register for the output.

The final example shown in Figure E.3 is a Verilog version of the simple computer model from Section 8.3 and Figure 8.7. In addition to Verilog features and statements seen in the earlier examples, two new Verilog statements appear. The Verilog wire statement is similar to an internal signal declaration and concurrent signal assignment in VHDL. A concurrent Verilog assign statement can also be used if the signal is already declared. The value of the signal on the right is copied or "wired" to the signal on the left. Note hierarchy is used when the memory for the computer is generated using an Altera LPM RAM function call in a component instantiation and the LPM RAM parameters are defined using Verilog's defparam statement.

VHDL Model of ALU

```
entity ALU is
  port(ALU_control   : in std_logic_vector(2 downto 0);
       Ainput, Binput: in std_logic_vector(15 downto 0);
       Clock         : in std_logic;
       Reg_output: out std_logic_vector(15 downto 0));
end ALU;

architecture RTL of ALU is
signal ALU_output: std_logic_vector(15 downto 0);
begin
 process (ALU_Control, Ainput, Binput)
  begin
   case ALU_Control(2 downto 1) is
    when "00" => ALU_output <= Ainput + Binput;
    when "01" => ALU_output <= Ainput - Binput;
    when "10" => ALU_output <= Ainput and Binput;
    when "11" => ALU_output <= Ainput or Binput;
    when others => ALU_output <=
                  "0000000000000000";
   end case;
  end process;
 process
 begin
  wait until rising_edge(Clock);
  if ALU_control(0) = '1' then
   Reg_output <= ALU_output(14 downto 0) & "0";
  else
   Reg_output <= ALU_output;
  end if;
 end process;
end RTL;
```

Verilog Model of ALU

```
module ALU ( ALU_control, Ainput, Binput,
             Clock, Reg_output);
 input [2:0] ALU_control;
 input [15:0] Ainput;
 input [15:0] Binput;
 input Clock;
 output[15:0] Reg_output;
 reg [15:0] Reg_output;
 reg [15:0] ALU_output;

always @(ALU_control or Ainput or Binput)

case (ALU_control[2:1])
   0: ALU_output = Ainput + Binput;
   1: ALU_output = Ainput - Binput;
   2: ALU_output = Ainput & Binput;
   3: ALU_output = Ainput | Binput;
   default: ALU_output = 0;
 endcase


always @(posedge Clock)
 if (ALU_control[0]==1)
   Reg_output = ALU_output << 1;
 else
   Reg_output = ALU_output;
endmodule
```

ALU Block Diagram

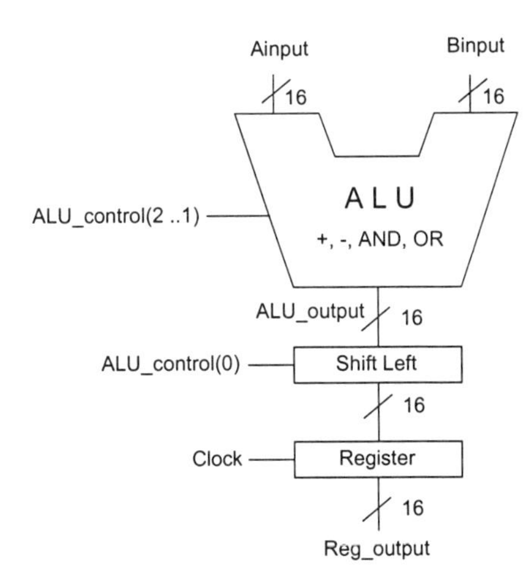

Figure E.2 Comparison of an ALU model in VHDL and Verilog

```
        //Simple Computer Design from Chapter 8 in Verilog
module SCOMP (clock,reset,program_counter,register_A,
                         memory_data_register_out, instruction_register);
  input clock,reset;
  output [7:0] program_counter;
  output [15:0] register_A, memory_data_register_out, instruction_register;

  reg  [15:0] register_A,  instruction_register;
  reg  [7:0] program_counter;
  reg  [3:0] state;
  reg [7:0] memory_address_register;
  reg memory_write;

              // State Encodings
parameter     reset_pc        = 0,
              fetch           = 1,
              decode          = 2,
              execute_add     = 3,
              execute_store   = 4,
              execute_store2  = 5,
              execute_store3  = 6,
              execute_load    = 7,
              execute_jump    = 8;
      wire [15:0] memory_data_register;
      wire [15:0] memory_data_register_out = memory_data_register;
      wire [15:0] memory_address_register_out = memory_address_register;
      wire memory_write_out = memory_write;

              // Use LPM function for computer's memory (256 16-bit words)
LPM_RAM_DQ  LPM_RAM_DQ_component (
                          .address (memory_address_register_out),
                          .inclock (clock),
                          .data (register_A),
                          .we (memory_write_out),
                          .q (memory_data_register));
      defparam
              LPM_RAM_DQ_component.LPM_WIDTH = 16,
              LPM_RAM_DQ_component.LPM_WIDTHAD = 8,
              LPM_RAM_DQ_component.LPM_INDATA = "REGISTERED",
              LPM_RAM_DQ_component.LPM_ADDRESS_CONTROL = "UNREGISTERED",
              LPM_RAM_DQ_component.LPM_OUTDATA = "UNREGISTERED",
              LPM_RAM_DQ_component.USE_EAB = "ON",
                    // Reads in mif file for initial program and data values
              LPM_RAM_DQ_component.LPM_FILE = "program.mif";

 always @(posedge clock or posedge reset)
  begin
   if (reset)
     state = reset_pc;
   else
     case (state)
                    // reset the computer, need to clear some registers
              reset_pc :
              begin
                          program_counter = 8'b00000000;
```

```
                memory_address_register = 8'b00000000;
                register_A = 16'b0000000000000000;
                memory_write = 0;
                state = fetch;
end

        // Fetch instruction from memory and add 1 to program counter
fetch :
begin
                instruction_register = memory_data_register;
                program_counter = program_counter + 1;
                memory_write = 0;
                state = decode;
end

        // Decode instruction and send out address of any data operands
decode :
begin
                memory_address_register = instruction_register[7:0];
                case (instruction_register[15:8])
                        8'b00000000:
                            state = execute_add;
                        8'b00000001:
                            state = execute_store;
                        8'b00000010:
                            state = execute_load;
                        8'b00000011:
                            state = execute_jump;
                        default:
                            state = fetch;
                endcase
end

        // Execute the ADD instruction
execute_add :
begin
                register_A = register_A + memory_data_register;
                memory_address_register = program_counter;
                state = fetch;
end

        // Execute the STORE instruction,needs 3 clock cycles for memory write)
execute_store :
begin
        // write register_A to memory
                memory_write = 1;
                state = execute_store2;
end
        // This state ensures that the memory address is valid until
        // after memory_write goes low
execute_store2 :
begin
                memory_write = 0;
                state = execute_store3;
end
```

```
            execute_store3 :
            begin
                        memory_address_register = program_counter;
                        state = fetch;
            end

                // Execute the LOAD instruction
            execute_load :
            begin
                        register_A = memory_data_register;
                        memory_address_register = program_counter;
                        state = fetch;
            end

                // Execute the JUMP instruction
            execute_jump :
            begin
                        memory_address_register = instruction_register[7:0];
                        program_counter = instruction_register[7:0];
                        state = fetch;
            end

                // Default case – an undefined opcode
            default :
            begin
                        memory_address_register = program_counter;
                        memory_write = 0;
                        state = fetch;
            end
        endcase
    end
endmodule
```

Figure E.3 Verilog model of the simple computer design from Chapter 8

Multiple Verilog module declarations can reside in a single file or in different files. Hierarchy is created in Verilog when higher-level modules invoke instances of lower-level modules and connect the port signals. Module declarations cannot be nested.

Sections 1.7 and 1.8 contain a tutorial on using the Verilog compiler. Numerous texts contain additional information on writing Verilog synthesis models. Some examples are *HDL Chip Design* by D. J. Smith, *Verilog HDL: A Guide to Digital Design and Synthesis* by S. Palnitkar, and *Verilog HDL Synthesis: A Practical Primer* by J. Bhasker.

Glossary

Assignment & Configuration File (.acf): An ASCII file (with the extension .acf) used by MAX+PLUS to store information about probe, pin, location, chip, clique, logic option, timing, connected pin and device assignments, as well as configuration settings for the Compiler, Simulator, and Timing Analyzer for an entire project. The ACF stores information entered with menu commands in all MAX+PLUS II applications. You can also edit an ACF manually in a Text Editor window.

Active-high (Active-low) node: A node that is activated when it is assigned a value one (zero) or Vcc (Gnd).

AHDL: Acronym for Altera Hardware Description Language. Design entry language that supports Boolean equation, state machine, conditional, and decode logic. It also provides access to all Altera and user-defined macrofunctions.

Ancillary file: A file that is associated with a MAX+PLUS II project, but is not a design file in the project hierarchy tree.

Antifuse: Any of the programmable interconnect technologies forming electrical connection between two circuit points rather than making open connections.

Architecture: Describes the behavior, RTL or dataflow, and/ or structure of a VHDL entity. An architecture is created with an architecture body. A single entity can have more than one architecture. In some VHDL tools, configuration declarations are used to specify which architectures to use for each entity.

Array: A collection of one or more elements of the same type that are accessed using one or more indices depending on dimension of array. Array data types are declared with an array range and array element type.

ASIC: Acronym for Application-Specific Integrated Circuit. A circuit whose final photographic mask process is user design dependent.

ASM: Acronym for Algorithmic State Machine Chart. A flow-chart based method used to represent a state diagram.

Assert: A statement that checks whether a specified condition is true. If the condition is not true, a report is generated during simulation.

Assignment: In VHDL, assignment refers to the transfer of a value to a symbolic name or group, usually through a Boolean equation. The value on the right side of an assignment statement is assigned to the symbolic name or group on the left.

Asynchronous input: An input signal that is not synchronized to the device Clock.

Attribute: A special identifier used to return or specify information about a named entity. Predefined attributes are prefixed with the ' character.

Back annotation: Process of incorporating time delay values into a design netlist reflecting the interconnect capacitance obtained from a completed design. Also, in Altera's case, the process of copying device and resource assignments made by the Compiler into the Assignment and Configuration File for a project. This process preserves the current fit in future compilations.

Block: A feature that allows partitioning of the design description within an architecture.

Buried node: A combinatorial or registered signal that does not drive an output pin.

Cell: A logic function. It may be a gate, a flip-flop, or some other structure. Usually, a cell is small compared to other circuit building blocks.

Cell library: The collective name for a set of logic functions defined by the manufacturer of an FPLD or ASIC. Simulation and synthesis tools use cell libraries when simulating and synthesizing a model.

CLB: Acronym for Configurable Logic Block. This element is the basic building block of the Xilinx FPGA product family.

Clock: A signal that triggers registers. In a flip-flop or state machine, the clock is an edge-sensitive signal. In edge-triggered flip-flops, the output of the flip-flop can change only on the clock edge.

Clock enable: The level-sensitive signal on a flip-flop with E suffix, e.g., DFFE. When the Clock enable is low, clock transitions on the clock input of the flip-flop are ignored.

Compiler Netlist File (.cnf): A binary file (with the extension .cnf) that contains the data from a design file. The CNF is created by the Compiler Netlist Extractor module of the MAX+PLUS II Compiler.

Component: Specifies the port of a primitive or macrofunction in VHDL. A component consists of the name of the primitive or macrofunction, and a list of its inputs and outputs. Components are specified in the Component declaration.

Component instantiation: A concurrent statement that references a declared component and creates one unique instance of that component.

Configuration EPROM: A serial EPROM designed to configure (program) a CPLD.

Concurrent statements: HDL statements that are executed in parallel.

Configuration: It maps instances of VHDL components to design entities and describes how design entities are combined to form a complete design. Configuration declarations are used to specify which architectures to use for each entity.

Configuration scheme: The method used to load configuration (programming) data into an FPGA.

Constant: An object that has a constant value and cannot be changed.

Control unit: The hardware of a machine that controls the data path.

CPLD: Acronym for complex programmable logic device. CPLDs include an array of functionally complete or universal logic cells with an interconnection network.

Data Path: The hardware path that provides data processing and transfer of information in a machine, as opposed to the controller.

Design entity: A file that contains description of the logic for a project and is compiled by the Compiler.

Design library: Stores VHDL units that have already been compiled. These units can be referenced in VHDL designs.

Design unit: A section of VHDL description that can be compiled separately. Each design unit must have a unique name within the project.

Dual-purpose pins: Pins used to configure an FPLD device that can be used as I/O pins after initialization.

Dynamic reconfigurability: Capability of an FPLD to change its function "on -the-fly"

Embedded Array Block (EAB): A physically grouped set of 8 embedded cells that implement memory (RAM or ROM) or combinatorial logic in a FLEX 10K device. A single EAB can implement a memory block of 256 x 8, 512 x 4, 1,024 x 2, or 2,048 x 1 bits.

EPLD: Acronym for EPROM programmable logic devices. This is a PLD that uses EPROM cells to internally configure the logic function. Also, erasable programmable logic device.

Event: The change of value of a signal. Usually refers to simulation.

Event scheduling: The process of scheduling of signal values to occur at some simulated time.

Excitation function: A Boolean function that specifies logic that directs state transitions in a state machine.

Exit condition: An expression that specifies a condition under which a loop should be terminated.

FLEX 10K and FLEX 10KA: An Altera device family based on Flexible Logic Element MatriX architecture. This SRAM-based family offers high-performance, register-intensive, high-gate-count devices with embedded arrays. The FLEX 10K device family includes the EPF10K100, EPF10K70, EPF10K50, EPF10K40, EPF10K30, EPF10K20, and EPF10K10 devices.

Fan-out: The number of output signals that can be driven by the output of a logic cell.

Fast Track interconnect: Dedicated connection paths that span the entire width and height of a FLEX device. These connection paths allow the signals to travel between all LABs in device.

Field name: An identifier that provides access to one element of a record data type.

File type: A data type used to represent an arbitrary-length sequence of values of a given type.

For loop: A loop construct in which an iteration scheme is a for statement.

Finite state machine: The model of a sequential circuit that cycles through a predefined sequence of states.

Fitting: Process of making a design fit into a specific FPLD architecture. Fitting involves technology mapping, placement, optimization, and partitioning among other operations.

Flip-flop: An edge-sensitive memory device (cell) that stores a single bit of data.

Floorplan: Physical arrangement of functions within a connection framework of signal routing channels.

FPGA: Acronym for field programmable gate array. A regular array of logic cells that is either functionally complete or universal with an interconnection network of signal routing channels.

FPLD: Acronym for field programmable logic device. An integrated circuit used for implementing digital hardware that allows the end user to configure the chip to realize different designs. Configuring such a device is done using either a special programming unit or by doing it " in system". FLPDs include both CPLDs and FPGAs.

Functional prototype: Specifies the ports of a primitive or macrofunction in AHDL. It consists of the name of the primitive or macrofunction, and a list of its inputs and outputs in exact order in which they are used. An instance of the primitive or macrofunction can be inserted with an Instance declaration or an in line reference.

Functional simulation: A simulation mode that simulates the logical performance of a project without timing information.

Functional test vector: The input stimulus used during simulation to verity a VHDL model operates functionally as intended.

Functionally complete: Property of some Boolean logic functions permitting them to make any logic function by using only that function. The properties include making the AND function with an invert or the OR function with an invert or the OR function with an invert.

Fuse: A metallic interconnect point that can be electrically changed from short circuit to an open circuit by applying electrical current.

Gate: An electronic structure, built from transistors that performs a basic logic function.

Gate array: Array of transistors interconnected to form gates. The gates in turn are configured to form larger functions.

Gated clock: A clock configuration in which the output of an AND or OR gate drives a clock.

Generic: A parameter passed to a VHDL entity, component or block that describes additional, instance-specific information about that entity, component or block.

Glitch or spike: A narrow output pulse that occurs when a logic level changes two or more times over a short period.

Global signal: A signal from a dedicated input pin that does not pass through the logic array before performing its specified function. Clock, Preset, Clear, and Output Enable signals can be global signals.

GND: A Low-level input voltage. It is the default inactive node value.

Graphic Design File (.gdf): A schematic design file (with the extension .gdf) created with the MAX+PLUS II Graphic Editor.

HDL: Acronym for Hardware Description Language. A special language used to describe digital hardware.

Hexadecimal: The base 16 number system (radix). Hexadecimal digits are 0 through 9 and A through F.

Hierarchy: The structure of a design description, expressed as a tree of related components.

Identifier: A sequence of characters that uniquely identify a named entity in a design description.

Index: A scalar value that specifies an element or range of elements within an array.

Input vectors: Time-ordered binary numbers representing input values sequences to a simulation program.

I/O cell register: A register on the periphery of a FLEX 8000 device or a fast input-type logic cell that is associated with an I/O pin.

IP core: An intellectual property (IP) core is a previously developed synthesizable hardware design that provides a widely used function. Commercially licensed IP cores include functions such as microprocessors, microcontrollers, bus interfaces, multimedia and DSP operations, and communications controllers.

LAB: Acronym for Logic Array Block. The LAB is the basic building block of the Altera MAX family. Each LAB contains at least one macrocell, an I/O block, and an expander product term array.

Latch: A level-sensitive clocked memory device (cell) that stores a single bit of data. A High to low transition on the Latch Enable signal fixes the contents of the latch at the value of the data input until the next Low-to-High transition on Latch Enable.

Latch enable: A level-sensitive signal that controls a latch. When it is High, the input flows through the output; when it is Low, the output holds its last value.

Library: In VHDL a library statement is used to store analyzed design units.

Literal: A value that can be applied to an object to some type.

Logic Synthesizer: The Compiler module that uses several algorithms to minimize gate count, remove redundant logic, and utilize the device architecture as efficiently as possible. Processing can be customized with logic options and logic synthesis style assignments. This module also applies logic synthesis techniques to help implement timing requirements for a project.

Least Significant Bit (LSB): The bit of a binary number that contributes the smallest quantity to the value of that number, i.e., the last member in a bus or group name. For example, the LSB for a bus or group named a[31..0] is a[0] (or a0).

Logic Cell (LC): The generic term for a basic building block of an Altera device. In MAX devices, a logic cell (also called a macrocell) consists of two parts: combinatorial logic and a configurable register. The combinatorial logic allows a wide variety of logic functions. In FLEX 10K devices, a logic cell (also called a logic element) consists of a look-up table (LUT) and a programmable register to support sequential functions.

Logic element: A basic building block of an Altera FLEX device. It consists of a look-up table i.e., a function generator that quickly computes any function of four variables, and a programmable flip-flop to support sequential functions.

LPM: Acronym for library of Parameterized Modules. Denotes Altera's library of design units that contain one or more changeable parts, and parameters that are used to customize a design unit as the application requires.

Macro: When used with FPGAs, a logic cell configuration that can be repeated as needed. It can be a Hard or a Soft macro. Hard macros force predefined place and route rules between logic cells.

Macrocell: In FPLDs, a portion of the FPLD that is smallest indivisible building block. In MAX devices it consists of two parts: combinatorial logic and a configurable register.

MAX: Acronym for Multiple Array MatriX, which is an Altera product family. It is usually considered to be a CPLD.

MAX+PLUS II: Acronym for multiple array matrix programmable logic user system II. A set of computer aided design (CAD) tools that allow design and implementation of custom logic circuits with Altera's MAX and FLEX CPLD devices.

Memory Initialization File (.mif): An ASCII file (with the extension .mif) used by MAX+PLUS II to specify the initial content of a memory block (RAM or ROM), i.e., the initial data values for each memory address. This file is used during project compilation and/or simulation.

Mealy state machine: A type of state machine in which the outputs are a function of the inputs and the current state.

Moore state machine: A state machine in which the present state depends only on its previous input and previous state, and the present output depends only on the present state. In general Moore states machines have more states than a Mealy machine.

Most Significant Bit (MSB): The bit of a binary number that contributes the greatest quantity to the value of that number, and the first member in a bus or group name. For example, the MSB for a bus named a[31..0] is a[31].

Mode: A direction of signal (either in, out, inout or buffer) used as subprogram parameter or port.

Model: A representation that behaves similarly to the operation of some digital circuit.

MPLD: Acronym for Mask Programmed Logic Device.

Netlist: A text file that describes a logic design. Minimal requirements are identification of function elements, inputs, outputs, and connections.

Netist synthesis: Process of deriving a netlist from an abstract representation, usually from a hardware description language.

NRE: Acronym for Non-Recurring Engineering expense. It reefers to one-time charge covering the use of design facilities, masks and overhead for test development of ASICs.

Object: A named entity of a specific type that can be assigned a value. Object in VHDL include signals, constants, variables and files.

Octal: The base 8 number system (radix). Octal digits are 0 though 7.

One Hot Encoding: A design technique used more with FPGAs than CPLDs. Only one flip-flop output is active at any time. One flip-flop per state is used. State outputs do not need to be decoded and they are hazard free.

Package: A collection of commonly used VHDL constructs that can be shared by more than one design unit.

PAL: Acronym for programmable array logic. A relatively small FPLD containing a programmable AND plane followed by a fixed-OR plane.

Parameter: An object or literal passed into a subprogram via that subprogram's parameter list.

Partitioning: Setting boundaries between subsections of a system or between multiple FPLD devices.

Physical types: A data type used to represent measurements.

Pin Number: A number used to assign an input or output signal in a design file, which corresponds to the pin number on an actual device.

PLA: (programmable logic array) a relatively small FPLD that contains two levels of programmable logic-an AND plane and an OR plane.

Placement: Physical assignment of a logical function to a specific location within an FPGA. Once the logic function is placed, its interconnection is made by routing.

PLD: Acronym for programmable logic device. This class of devices comprise PALs, PLAs, FPGAs, and CPLDs.

Port: A symbolic name that represents an input or output of a primitive or of a macrofunction design file.

Primitive: One of the basic functional blocks used to design circuits with MAX+PLUS II software. Primitives include buffers, flip-flops, latch, logical operators, ports, etc.

Process: A basic VHDL concurrent statement represented by a collection of sequential statements that are executed whenever there is an event on any signal that appears in the process sensitivity list, or whenever an event occurs that satisfies condition of a wait statement within the process.

Product Term: Two or more factors in a Boolean expression combined with an AND operator constitute a product term, where "product" means "logic product."

Programmable switch: A user programmable switch that can connect a logic element or input/output element to an interconnect wire or one interconnect wire to another.

Project: A project consists of all files that are associated with a particular design, including all subdesign files and ancillary files created by the user or by MAX+PLUS II software. The project name is the same as the name of the top-level design file without an extension.

Propagation delay: The time required for any signal transition to travel between pins and/or nodes in a device.

Radix: A number base. Group logic level and numerical values are entered and displayed in binary, decimal, hexadecimal, or octal radix in MAX+PLUS II.

Reset: An active-high input signal that asynchronously resets the output of a register to a logic Low (0) or a state machine to its initial state, regardless of other inputs.

Range: A subset of the possible values of a scalar type.

Register: A memory device that contains several latches or flip-flops that are clocked from the same clock signal.

Resource: A resource is a portion of a device that performs a specific, user-defined task (e.g., pins, logic cells, interconnection network).

Retargetting: A process of translating a design from one FPGA or other technology to another. Retargetting involves technology-mapping optimization.

Reset: An active-high input signal that asynchronously resets the output of a register to a logic Low (0) or a state machine to its initial state, regardless of other inputs.

Ripple Clock: A clocking scheme in which the Q output of one flip-flop drives the Clock input to another flip-flop. Ripple clocks can cause timing problems in complex designs.

RTL: Acronym for Register Transfer Level. The model of circuit described in VHDL that infers memory devices to store results of processing or data transfers. Sometimes it is referred to as a dataflow-style model.

Scalar: A data type that has a distinct order of its values, allowing two objects or literals of that type to be compared using relational operators.

Semicustom: General category of integrated circuits that can be configured directly by the user of an IC. It includes gate arrays and FPLD devices.

Signal: In VHDL a data object that has a current value and scheduled future values at simulation times. In RTL models signals denote direct hardware connections.

Simulation: Process of modeling a logical design and its stimuli in which the simulator calculates output signal values.

Slew rate: Time rate of change of voltage. Some FPGAs permit a fast or slow slew rate to be programmed for an output pin.

Slice: A one-dimensional, contiguous array created as a result of constraining a larger one-dimensional array.

Speed performance: The maximum speed of a circuit implemented in an FPLD. It is set by the longest delay through any for combinational circuits, and by maximum clock frequency at which the circuit operates properly for sequential circuits.

State transition diagram: A graphical representation of the operation of a finite state machine using directed graphs.

State: A state is implemented in a device as a pattern of 1's and 0's (bits) that are the outputs of multiple flip-flops (collectively called a state machine state register).

Structural-type architecture: The level at which VHDL describes a circuit as an arrangement of interconnected components.

Subprogram: A function or procedure. It can be declared globally or locally.

Sum-of-products: A Boolean expression is said to be in sum-of-products form if it consists of product terms combined with the OR operator.

Synthesis: The process of converting the model of a design described in VHDL from one level of abstraction to another, lower and more detailed level that can be implemented in hardware.

Technology mapping: Process of translating the function of a design from one technology to another. All versions of the design would have the same function, but the logic cells used would be very different.

Test bench: A VHDL model used to verify the correct behavior of another VHDL model, commonly known as the unit under test.

Tri-state Buffer: A buffer with an input, output, and controlling Output Enable signal. If the Output Enable input is High, the output signal equals the input. If the Output Enable input is Low, the output signal is in a state of high impedance. Tri-state outputs can be tied together but only one should ever be enabled at any given time.

Timing Simulation: A simulation that includes the actual device delay times.

Two's Complement: A system of representing binary numbers in which the negative of a number is equal to its logic inverse plus 1. In VHDL, you must declare a two's complement binary number with a signed data type or use the signed library.

Type: A declared name and its corresponding set of declared values representing the possible values the type.

Type declaration: A VHDL declaration statement that creates a new data type. A type declaration must include a type name and a description of the entire set of possible values for that type.

Universal logic cell: A logic cell capable of forming any combinational logic function of the number of inputs to the cell. RAM, ROM and multiplexers have been used to form universal logic cells. Sometimes they are also called look-up tables or function generators.

Usable gates: Term used to denote the fact that not all gates on an FPGA may be accessible and used for application purposes.

Variable: In VHDL, a data object that has only current value that can be changed in variable assignment statement.

Verilog: An HDL with features similar to VHDL with a syntax reminiscent of C.

VCC: A high-level input voltage represented as a High (1) logic level in binary group values.

VHDL: Acronym for VHSIC (Very High Speed Integrated Circuits) Hardware Description Language. VHDL is used to describe function, interconnect and modeling.

VITAL: Acronym for VHDL Initiative Toward ASIC Libraries. An industry-standard format for VHDL simulation libraries.

Index

O

P

R

S

T

U

V

W

X

About the Accompanying CD-ROM

Rapid Prototyping of Digital Systems, Second Edition, includes a CD-ROM that contains both Altera's MAX+PLUS II Student Edition Version 10.1 and all of the example designs from the text.

MAX+PLUS II Software

The companion CD-ROM contains Altera's MAX+PLUS II Student Edition programmable logic development software. The MAX+PLUS II software is a fully integrated design environment that offers unmatched flexibility and performance. The intuitive graphical interface is complemented by complete and instantly accessible online documentation, which makes learning and using the MAX+PLUS II software quick and easy. The MAX+PLUS II version 10.1 Student Edition software offers the following features:

- Operates on stand-alone PCs running Windows 98, Windows NT, and Windows 2000
- Design entry includes graphical and text-based, including VHDL, Verilog HDL, and the Altera Hardware Description Language (AHDL),
- Design compilation for product-term and look-up table device architectures, with device support for members of the Altera® FLEX® 10K, FLEX 8000, FLEX 6000, MAX® 7000, and MAX 7000S device families
- Design verification with full timing and functional simulation

Installing the MAX+PLUS II Software

Insert the MAX+PLUS II CD-ROM in your CD-ROM drive. In the Start menu, choose RUN and type **CD-ROMdrive:\setup.exe** in the Command Line box. Be sure to follow the instructions in the SE_READ.txt file. The instructions in SE_READ.txt will guide you through the installation and the online web registration procedure required to obtain a student version authorization code from Altera via email. Browse the file, **index.htm**, on the CD-ROM using a web browser for additional information and click on the link to the book's website for any software updates.

Design Examples from the Book

Design examples from the book are located in the CD-ROM directories, booksoft/chap*x*, where *x* is the chapter number. To use the design files, copy them to the hard disk drive to your max2work directory or a subdirectory. In addition to *.gdf and *.vhd design files, be sure to copy any *.acf, *.scf, or *.hif files for each project. The UP1core library files are in booksoft/chap5. Users with a **UP 1X** or **UP 2** board will need to assign the device type to a FLEX 10K70 on each CD-ROM design and recompile – See Section 1.1.

In Windows, all CD-ROM files have the read only attribute set. After copying CD-ROM files with Windows, you will need to clear the read only attribute on the hard disk copy of the files to be able to use them with the Altera tools. Using Windows Explorer, **select the files** and then use **FILE => PROPERTY** to **clear the read only attribute**. Control A can be used to select all of the files in a directory in Explorer. The DOS copy command automatically clears the read only attribute. Failure to reset the read only attribute will cause file errors when running the Altera tools.

This CD-ROM is distributed by Kluwer Academic Publishers with ***ABSOLUTELY NO SUPPORT*** and ***NO WARRANTY*** from Kluwer Academic Publishers. Kluwer Academic Publishers and the authors shall not be liable for damages in connection with, or arising out of, the furnishing, performance or use of the CD-ROM.

FOR ALTERA USE ONLY
Approved: ____________
Order date: ____________

Design Laboratory Package Order Form (Photocopy or Tear Out)

Contact name: ______________________ ☐ Faculty ☐ Staff ☐ Student ☐ Other ________

Institution name: ______________________ Department: ______________________

Address: ____

City: ______________ State: __________ ZIP: _________ Country: __________

Phone number: ______________________ FAX number: ______________________

Email: ______________________ WWW site: http://______________________

Order(s) will be shipped to the above address.

The UP2 Design Laboratory Package contains the UP2 Education Board populated with EPF10K70RC240-4 and EPM7128SLC84-7 devices, MAX+PLUS II Version 10.1 Student Edition, Download / Programming Cable, and the Design Laboratory Package User Guide (Retail Value $975).

ITEM	DESCRIPTION	QTY	UNIT PRICE	TOTAL
UP-2-DLP-70	UP2 Design Laboratory Package		$149.00	
	Shipping Charges			-0-
			TOTAL	

PAYMENT METHOD: ___ CHECK ___ MONEY ORDER ___ CREDIT CARD

All orders must be accompanied with payment.

Credit Card: _______MasterCard _______VISA

Credit Card #:______________________ Expiration Date: ____________

Signature: ______________________ Date:______________

Send to: **Ralene Marcoccia – University Program Coordinator**
Altera Corporation
101 Innovation Drive
San Jose, CA 95134
Tel: (408) 544-7198
FAX: (408) 544-6401